Individual and Occupational Determinants

Occupational Safety, Health, and Ergonomics: Theory and Practice

Series Editor: Danuta Koradecka

(Central Institute for Labour Protection – National Research Institute)

This series will contain monographs, references, and professional books on a compendium of knowledge in the interdisciplinary area of environmental engineering, which covers ergonomics and safety and the protection of human health in the working environment. Its aim consists in an interdisciplinary, comprehensive and modern approach to hazards, not only those already present in the working environment, but also those related to the expected changes in new technologies and work organizations. The series aims to acquaint both researchers and practitioners with the latest research in occupational safety and ergonomics. The public, who want to improve their own or their family's safety, and the protection of heath will find it helpful, too. Thus, individual books in this series present both a scientific approach to problems and suggest practical solutions; they are offered in response to the actual needs of companies, enterprises, and institutions.

Individual and Occupational Determinants: Work Ability in People with Health Problems
Joanna Bugajska, Teresa Makowiec-Dąbrowska, Tomasz Kostka

Healthy Worker and Healthy Organization: A Resource-Based Approach
Dorota Żołnierczyk-Zreda

Emotional Labour in Work with Patients and Clients: Effects and Recommendations for Recovery
Dorota Żołnierczyk-Zreda

New Opportunities and Challenges in Occupational Safety and Health Management
Daniel Podgórski

Emerging Chemical Risks in the Work Environment
Małgorzata Pośniak

Visual and Non-Visual Effects of Light: Working Environment and Well-Being
Agnieszka Wolska, Dariusz Sawicki, Małgorzata Tafil-Klawe

Occupational Noise and Workplace Acoustics: Advances in Measurement and Assessment Techniques
Dariusz Pleban

Virtual Reality and Virtual Environments: A Tool for Improving Occupational Safety and Health
Andrzej Grabowski

Head, Eye, and Face Personal Protective Equipment: New Trends, Practice and Applications
Katarzyna Majchrzycka

Nanoaerosols, Air Filtering and Respiratory Protection: Science and Practice
Katarzyna Majchrzycka

Microbial Corrosion of Buildings: A Guide to Detection, Health Hazards, and Mitigation
Rafał L. Górny

Respiratory Protection Against Hazardous Biological Agents
Katarzyna Majchrzycka, Justyna Szulc, Małgorzata Okrasa

For more information about this series, please visit: https://www.crcpress.com/Occupational-Safety-Health-and-Ergonomics-Theory-and-Practice/book-series/CRCOSHETP

Individual and Occupational Determinants

Work Ability in People with Health Problems

Edited by
Joanna Bugajska, Teresa Makowiec-Dąbrowska,
and Tomasz Kostka

CRC Press
Taylor & Francis Group
Boca Raton London New York

CRC Press is an imprint of the
Taylor & Francis Group, an **informa** business

First edition published 2021
by CRC Press
6000 Broken Sound Parkway NW, Suite 300, Boca Raton, FL 33487-2742

and by CRC Press
2 Park Square, Milton Park, Abingdon, Oxon, OX14 4RN

Library of Congress Cataloging-in-Publication Data
Names: Bugajska, Joanna, editor. | Makowiec-Dąbrowska, Teresa, editor. |
Kostka, Tomasz, editor.
Title: Individual and occupational determinants : work ability in people
with health problems / edited by Joanna Bugajska, Teresa
Makowiec-Dąbrowska, Tomasz Kostka.
Description: First edition. | Boca Raton : CRC Press, 2020. |
Series: Occupational safety, health, and ergonomics | Includes bibliographical
references and index.
Identifiers: LCCN 2020025542 (print) | LCCN 2020025543 (ebook) |
ISBN 9780367469337 (hardback) | ISBN 9781003088479 (ebook)
Subjects: LCSH: Disability evaluation. | Disability evaluation—Case
studies. | Work capacity evaluation.
Classification: LCC RC963.4 .I53 2020 (print) | LCC RC963.4 (ebook) |
DDC 616.07/5—dc23
LC record available at https://lccn.loc.gov/2020025542
LC ebook record available at https://lccn.loc.gov/2020025543

ISBN: 9780367469337 (hbk)
ISBN: 9780367542795 (pbk)
ISBN: 9781003088479 (ebk)

Typeset in Times
by codeMantra

Contents

PART I   General Issues

PART II   Work Ability and Age

PART III   Work Ability and Chronic Diseases

PART IV   Work Ability and Disabilities

Preface

Polish scientist, philosopher and naturalist Wojciech Jastrzębowski in 1857 in his article "An Outline of Ergonomics, or the Science of Work" (Jastrzębowski 1997) has written:

"And apart from this, our forces and abilities, whereby and trough the guidance of which we perform our work, develop through our exercise of work, perfecting in the proper respect, and thereby contributing to the advancement and perfection of our entire being, which is the condition for our felicity and without which our existence is meagre and ever under the threat of doom. For it is well-known that our vital forces grow weak and impoverished as much by the lack of their exercise as by their abuse; and that they are maintained in their popper condition, growing and increasing by their proper and moderated exercise, which we call work, and whereby we improve things, people, and ourselves, making them and us more conductive to the service of the common good."

Citing the words of Wojciech Jastrzębowski, the creator of the term "ergonomics", we would like to emphasize the multidimensionality of the quoted statement. Despite the fact that the text was written in mid-19th century, the statements regarding the importance of balanced activities, including professional activities, for human development and happiness, are still up to date. Moreover, Jastrzębowski emphasizes the preventive effect of physical activity on human health and well-being.

These words are very important nowadays, when we face, on the one hand, ageing of the population including occupationally active people, the increased prevalence of chronic diseases and the low employment rate in people with disabilities and, on the other hand, new working environment hazards resulting from the implementation of the Fourth Industrial Revolution (Industrie 4.0).

In this context, an adequate assessment of work ability, considering both the medical aspects and the work content and working environment, is of extreme importance.

This publication will address a very important problem related to current demographic trends in European Union countries and the resulting increase in the population of professionally active people of advanced age, people with chronic health issues and people with disabilities. The proper functioning of these people in their lives, including occupational activity, is only possible if they are provided with appropriate psychosocial and physical working conditions.

The publication will present the individual and occupational factors having the most remarkable effect on maintaining work ability in older people, and persons with chronic health issues and disabilities. The concept of work ability developed at the end of the 20th century in Finland has been accepted as the basis for discussions (Tuomi et al. 1997, Ilmarinen 2004). The theoretical aspects of work ability in older people and persons with health concerns are extended to include presentation of the results of research on work ability measured using Work Ability Index among Polish workers (Tuomi et al. 1998).

The publication will also present the concept of applying the International Classification of Functioning, Disability and Health (ICF) in order to assess the work ability of people with disabilities (WHO 2001).

Practical activities aimed at ensuring the possibility of performing professional work by people of advanced age and people with chronic diseases and disabilities will also be discussed.

REFERENCES

Ilmarinen, J., and S. Lehtinen, eds. 2004. *Past, Present, and Future of Work Ability: Proceedings of the 1st International Symposium on Work Ability*, 5–6 September 2001, Tampere, Finland. Helsinki: Finnish Institute of Occupational Health.

Info: https://locatorplus.gov/cgi-bin/Pwebrecon.cgi?DB=local&CNT=25&HIST=1&BOOL1 =as+a+phrase&FLD1=ISBN+(ISBN)&SAB1=9789518025811

Jastrzębowski, W. 1997. Rys ergonomji czyli nauki o pracy opartej na prawdach poczerpniętych z Nauki Przyrody. Outline of ergonomics, or the science of work based upon the truths drawn from the Science of Nature, transl. T. Bałuk-Ulewiczowa. Warszawa: Centralny Instytut Ochrony Pracy.

Tuomi, K., J. M. Ilmarinen, M. Klockars et al. 1997. Finnish research project on aging workers in 1981–1992. *Scand J Work Environ Health* 23(Suppl 1):7–11.

Tuomi, K., J. Ilmarinen, A. Jahkola, L. Katajarinne, and A. Tulkki. 1998. *Work Ability Index*. 2nd ed. Helsinki: Finnish Institute of Occupational Health.

WHO [World Health Organization]. 2001. International Classification of Functioning, Disability and Health (ICF). https://www.who.int/classifications/icf/en/ (accessed September 18, 2019).

Acknowledgment

Wishing a special thank you to Jolanta Schönhof-Wilkans for translating this book into English.

Editors

Joanna Bugajska, Ph.D., D.Med.Sc., is a medical specialist in internal medicine and occupational medicine, and a Professor and Head of Department of Ergonomics at the Central Institute for Labour Protection – National Research Institute, Warsaw, Poland. Her scientific, publication and expertise activities include the following:

• Assessment of workload and the health effects of excessive loads
• Type of work, the method of its performance and the occurrence of musculoskeletal overload disorders
• Psychophysical aspects of work ability
• Principles of shaping work requirements to the psychophysical capabilities of the worker
• Impact of lifestyle, including physical professional and non-professional activity on health, quality of life and work ability
• Application of ICF to assess work ability and professional competences of people with disabilities
• Adapting the working environment to the needs of people with disabilities.

Joanna Bugajska is a member of editorial board and the author (co-author) of approximately 120 scientific publications in peer-reviewed journals: *International Journal of Occupational Safety and Ergonomics, Disabilities – Issues, Problems, Solutions*, and *Principles and Methods of Assessing the Working Environment*.

Teresa Makowiec-Dąbrowska, Ph.D., D.Med.Sc., is a long-term employee of the Nofer Institute of Occupational Medicine, Łódź, Poland. She has nearly 50 years of experience in the field of work physiology. Her scientific, publication and expertise activities include the following:

• The impact of the circadian rhythm on physiological functions and work ability
• Assessment of workload and the health effects of excessive loads
• Impact of the method of work performance on the locomotor system

- Assessment of physiological capabilities related to the requirements of the profession – work ability assessment
- Principles of shaping loads for employees of changed capabilities
- Impact of lifestyle, including physical professional and non-professional activity on health, quality of life and work ability
- Prevalence and effects of smoking.

Teresa Makowiec-Dąbrowska is the author (co-author) of approximately 200 scientific publications (including 92 publications indexed in international scientific databases) and has delivered many speeches at national and international scientific conferences.

Tomasz Kostka, Ph.D., D.Med.Sc., Prof., is a medical specialist in internal medicine, sports medicine, geriatrics and medical rehabilitation, and a Professor and Head of Department of Geriatrics, Medical University of Łódź, Poland. He is a Vice Rector for Teaching of the Medical University of Łódź and National Consultant in Geriatric Medicine in Poland.

His scientific, publication and expertise activities include the following:

- Gerontology and geriatrics, internal medicine, sports medicine, preventive medicine and rehabilitation
- Physiological and metabolic effects of physical activity in the elderly
- Muscle function, nutrition and antioxidant system
- Disability, sarcopenia, frailty and quality of life in older people.

Tomasz Kostka is a member of the editorial board and the author of more than 100 publications in peer-reviewed journals and reviewer of many scientific journals: *Current Gerontology and Geriatrics Research, European Geriatric Medicine, Journal of Health Policy, Insurance and Management, Journal of Aging Science* and *Frontiers in Geriatric Medicine.*

Contributors

Łukasz Baka, Ph.D., D.Sc., is a researcher at the Laboratory of Social Psychology, Central Institute for Labour Protection – National Research Institute, Warsaw, Poland. His research area involves work-related stress and counterproductive behaviours.

Katarzyna Hildt-Ciupińska, Ph.D., is a researcher in the Department of Ergonomics, Central Institute for Labour Protection – National Research Institute, Warsaw, Poland, and is responsible for research and development in the area of work–life balance, health and safety at work among employees, health promotion, healthy lifestyle and health behaviours of employees, as well as among older people and people with disability.

Łukasz Kapica, M.A., is a researcher at the Laboratory of Social Psychology, Central Institute for Labour Protection – National Research Institute, Warsaw, Poland. His research area focuses on traffic psychology and work and organizational psychology with particular emphasis on job crafting and job attitudes.

Joanna Kostka, Ph.D., D.Sc., is a researcher and academic teacher in the Department of Neurological Rehabilitation, Medical University of Łódź, Poland. She is a member of Committee of Rehabilitation, Physical Education and Social Integration of the Polish Academy of Science. Her research area is focused on physical activity, rehabilitation and functional performance of elderly and people with disabilities.

Elżbieta Łastowiecka-Moras, M.D.-Ph.D., is a researcher in the Department of Ergonomics, Laboratory of Physiology and Hygiene of Work, Central Institute for Labour Protection – National Research Institute, Warsaw, Poland. Her research area focuses on relationship between working conditions and physiological parameters of workers, assessment of workload, work ability and ageing.

Marzena Malińska, M.A., is a researcher in the Department of Ergonomics, Central Institute for Labour Protection – National Research Institute, Warsaw, Poland, and is responsible for research and development in the area of musculoskeletal disorders, work ability, health promotion at workplace, assessment of workload and the health effects of excessive loads.

Andrzej Najmiec, M.A., is a researcher at the Laboratory of Social Psychology, Central Institute for Labour Protection – National Research Institute, Warsaw, Poland. His research area is focused on psychosocial working conditions with special interest in stress and social support, work and organizational psychology, transport psychology and safety culture.

Karol Pawlak, M.A., is a researcher at the Polish International Classification of Functioning, Disability and Health (ICF) Council, Warsaw, Poland. His research area is focused on biopsychosocial model of disability, rehabilitation and employment of people with disabilities.

Karolina Pawłowska-Cyprysiak, M.A., is a researcher and assistant in the Department of Ergonomics, Central Institute for Labour Protection – National Research Institute, Warsaw, Poland. Her main activities include psychosocial and health determinants of professional and social activity among people with disabilities, work–life balance, learning and willingness to learn among older workers.

Halina Sienkiewicz-Jarosz, Ph.D., D.Med.Sc., Prof., is a member of the Main Board of Polish Neurological Society. Her current scientific interests focus on epidemiology of stroke, epilepsy, degenerative diseases of the central nervous system and mental disorders. She is involved in projects regarding psychosocial aspects of epilepsy, i.e. anxiety and depressive symptoms prevalence, stigmatization and psychological factors of drug compliance. She is the author of over 200 publications in scientific journals and monographs. She was an expert of the Ministry of Science and Information Technology for Specific Support Action NEURON-ERANET. She was also an expert of the Ministry of Health in the project Healthcare Needs Maps.

Maria Widerszal-Bazyl, Ph.D., D.Sc., is a researcher at the Laboratory of Social Psychology, Central Institute for Labour Protection – National Research Institute, Warsaw, Poland. Her research area has focused on psychosocial and personal causes of work-related stress, stress mechanisms, its health consequences and psychosocial risk management in organization.

Series Editor

Professor Danuta Koradecka, Ph.D., D.Med.Sc., is the Director of the Central Institute for Labour Protection – National Research Institute (CIOP-PIB) and a specialist in occupational health. Her research interests include the human health effects of hand-transmitted vibration; ergonomics research on the human body's response to the combined effects of vibration, noise, low temperature and static load; assessment of static and dynamic physical load; and development of hygienic standards as well as development and implementation of ergonomic solutions to improve working conditions in accordance with International Labour Organisation (ILO) convention and European Union (EU) directives. She is the author of more than 200 scientific publications and several books on occupational safety and health.

***

The "Occupational Safety, Health, and Ergonomics: Theory and Practice" series of monographs is focused on the challenges of the 21st century in this area of knowledge. These challenges address diverse risks in the working environment of chemical (including carcinogens, mutagens, endocrine agents), biological (bacteria, viruses), physical (noise, electromagnetic radiation) and psychophysical (stress) nature. Humans have been in contact with all these risks for thousands of years. Initially, their intensity was lower, but over time, it has gradually increased and now too often exceeds the limits of man's ability to adapt. Moreover, risks to human safety and health, so far assigned to the working environment, are now also increasingly emerging in the living environment. With the globalization of production and merging of labour markets, the practical use of the knowledge on occupational safety, health and ergonomics should be comparable between countries. The presented series will contribute to this process.

The Central Institute for Labour Protection – National Research Institute, conducting research in the discipline of environmental engineering, in the area of working environment and implementing its results, has summarized the achievements – including its own – in this field from 2011 to 2019. Such work would not be possible without cooperation with scientists from other Polish and foreign institutions as authors or reviewers of this series. I would like to express my gratitude to all of them for their work.

It would not be feasible to publish this series without the professionalism of the specialists from the Publishing Division, the Centre for Scientific Information and Documentation, and the International Cooperation Division of our Institute. The challenge was also the editorial compilation of the series and ensuring the efficiency of this publishing process, for which I would like to thank the entire editorial team of CRC Press – Taylor & Francis.

***

This monograph, published in 2020, has been based on the results of a research task carried out within the scope of the second to fourth stage of the Polish National Programme "Improvement of safety and working conditions" partly supported – within the scope of research and development – by the Ministry of Science and Higher Education/National Centre for Research and Development, and – within the scope of state services – by the Ministry of Family, Labour and Social Policy. The Central Institute for Labour Protection – National Research Institute is the Programme's main coordinator and contractor.

Part I

General Issues

1 Work Ability Assessment

An Important Element of Workers Health Protection

Joanna Bugajska
Central Institute for Labour Protection –
National Research Institute

Teresa Makowiec-Dąbrowska
Nofer Institute of Occupational Medicine

Work ability assessment is one of the key elements of workers' health protection. It belongs to the functions of occupational health service (OHS), focused on health protection in the working population against adverse conditions of working environment, the manner the occupational activity is implemented and preventive measures including workers health control. Systematic control of workers' health is also aimed at active employers' contribution to improvement of working conditions in order to reduce occupational hazards. It comprises procedures and studies involving workers' health assessment, oriented to identification of the work-related health disorders.

Work ability is a wide-ranging term, defined in different ways, depending on the purpose of definition and the accepted concept. However, regardless of the purpose, each case requires determining whether a given person is able to work. This ability is the result of interaction between the person, the task and the working environment. This means that work ability should not be determined without a detailed assessment of work-related tasks and working environment (Tengland 2011). Therefore, we should consider not only characteristics of a future worker including his/her physical and mental potential, knowledge and skills, but also job requirements and working conditions. According to one of the conceptions pertaining to the discussed issue, "work ability assessment is mainly based on medical criteria and understood as fitness for work" which, according to the definition, is the evaluation of worker's capacity to work, entailing no hazard for the worker and other persons (Cox et al. 2000; Serra et al. 2007). The criteria of fitness for work include determining the worker's ability to work and the risk associated with his or her workplace, considering all the ethical, economical and legal aspects of this issue. It is also emphasized that the basic goal of the future worker's health assessment, as part of the process of his or her work ability rating, is not exclusion from work but ensuring that a given person is able to effectively perform occupational tasks without risking his/her own and other people's safety and health. Another goal of work ability assessment is to determine the necessary and justified modifications or adjustments of psychosocial and physical working conditions.

A different concept of work ability assessment was proposed in the 1980s of the 20th century and next verified by Finnish researchers (Tuomi et al. 1998; Ilmarinen and Tuomi 2004). According to this concept, work ability is considered to be a four-level model, often presented as a habitable building. The first level incorporates functional capacity based on health status, and physical and mental fitness ensuring adequate functioning; the second level comprises knowledge, abilities and skills acquired so far (experience); the third level refers to motivation, attitudes and human values, whereas the fourth level includes work environment (work content, organization, psychosocial and physical demands and relationships between the management and workers). From the perspective of human life, all the levels included in the model are subject to changes. The main reasons for these changes include progressing ageing process, diseases, changes in motivation and values, changes in occupational skills (as they become outdated) and changes in the working environment resulting from, among other things, the progress in technology, means of production and communication (media).

Analysing the two concepts of work ability, we may conclude that these are complementary concepts, differing only in measurement approaches. Fitness for work is evaluated on the basis of objective assessment of health status and coherence between certain (also objectively stated) physiological and/or psychological characteristics of a future worker, and the requirements related to a particular job. According to Ilmarinen (2006), work ability is the workers' resources in relation to work demands. A worker's resources consist of health and ability, education and skills, and values and attitudes. Work covers the work environment and community, as well as the actual contents, demands and organization of work.

Both presented concepts indicate that work ability is a balance between, on the one hand, job characteristics (work content, psychosocial and physical job demands) and, on the other hand, the individual factors allowing workers to perform a given work, including mainly their psychophysical potential and competence. The balance can thus be disturbed by both the changes in the human being and the factors related to work environment. We constantly witness such changes.

According to European Agency for Safety and Health at Work (EU-OSHA), three main changes have been recently observed in global working environment: demographic changes, intensive globalization and technological changes. The aforementioned changes, related not only to ageing of the workforce but also to the new forms of employment as well as increased job insecurity and intensity, markedly affect the worker's psychophysical workload and well-being (Milczarek et al. 2012; EU-OSHA 2013). At the same time, the physical, psychological, qualitative and quantitative job demands towards both high skilled and unskilled workers are increasing. Given the increasingly complex situation on the labour market, this interaction between the factors connected with occupational processes and the individual perception of job demands is of key importance for work ability and in situations requiring coping with occupational stress. The industrial development in the developed countries has recently been dominated by the implementation of the so-called industrial revolution (Industrie 4.0). This fact is indicative of the widespread digitalization of production processes, the integration of all stages of production using information technology (IT) and also of the introduction of changes in manufacturing processes. Implementation of these ideas is closely connected with the changes in the implementation of new forms of employment and the impact of working environment on workers' functioning and health.

In the report of the European Foundation for the Improvement of Living and Working Conditions (Eurofound 2015), a plethora of new forms of employment have been named; the forms of employment that will be increasingly popular in the modern world include the following: employee sharing, job sharing, interim management, casual work (job on-call), voucher-based work, portfolio work, crowdsourcing, collaborative employment, freelancers and remote work based on information and telecommunication technologies (ICT).

The changes in working conditions and employment forms entail new psychosocial risks for the worker, such as increased job insecurity, higher employment competition, work under time pressure, decreased social support, "apparent" worker autonomy (tied autonomy) and strong dependency on the customer and partner requirements.

The above-mentioned changes in forms of employment and psychosocial risk factors bring about a plenty of adverse consequences, including mental workload, cognitive load, disturbed balance between work and private life, chronic excessive activation and problems with concentration and rest (Madsen et al. 2017). Such loads generate fatigue and occupational burnout, and finally severe depressive disorders resulting in long-term absenteeism and increasingly frequent social exclusion. The consequences of the changing demands towards the worker also include increasing use of substance by workers, stimulating their functioning at work and in private life and bringing about adverse health-related and social effects.

With the changes in working environment, quickly progressing changes in the occupationally active population are noted. The first changes are related to ageing of the society, involving prolonged occupational activity and the increased share of older people in occupationally active population. The second ones, also partly related to ageing of the society, involve the increase in the population of workers with chronic diseases. This phenomenon is mainly due to the prolonged periods of occupational activity and the increasing share of older persons and those with chronic diseases in the working population. With age, an increased prevalence of numerous diseases is noted, especially cardiovascular diseases, breathing problems, musculoskeletal disorders and mental illnesses, as well as hormonal and metabolic disorders, which are undoubtedly one of the most important reasons for work ability decrease. The occupational activity among people with disabilities is also low. According to the Survey of Economic Activity of People (BAEL), the employment index among the economically active population of working age for 2019 in Poland was 27.7% (GUS 2020).

These data indicate a high, unused potential of older people and persons with disabilities, but also the difficulties in occupational activation of these groups of people, often unskilled ones, especially as regards the development of new technologies. In this context, work ability assessment, especially in older people with chronic diseases and people with disabilities, is the basic element of workers' health protection.

REFERENCES

Cox, R. A. F., F. C. Edwards, and K. Palmer. 2000. *Fitness for work. The medical aspects.* 3rd ed. Oxford: Oxford Medical Publications.

EU-OSHA. 2013. *Priorities for occupational safety and health research in Europe: 2013– 2020.* Luxembourg: Publications Office of the European Union. https://osha.europa. eu/en/publications/reports/priorities-for-occupational-safety-and-health-research-in- europe-2013-2020/view. (accessed 28 January 2020).

Eurofound. 2015. *New forms of employment.* Luxembourg: Publications Office of the European Union.

GUS, BAEL (quaterly data), based on BON MRPiPS. 2020. http://niepelnosprawni.hostlab. pl/p,81,bael. (accessed 27 January 2020).

Ilmarinen, J. 2006. *Towards a longer worklife: Ageing and the quality of worklife in the European Union.* Helsinki: Finnish Institute of Occupational Health, Ministry of Social Affairs and Health.

Ilmarinen, J., and K. Tuomi. 2004. Past, present and future of work ability. *Proceedings of the 1st International Symposium on Work Ability*, 5–6 September 2001. Edited by: Ilmarinen, J., and Lehtinen, S. Book Series: People and Work: Research Reports. Vol. 65:1–25.

Madsen, I. E. H., S. T. Nyberg, L. L. Magnusson Hanson, J. E. Ferrie, K. Ahola, and IPD-Work Consortium. 2017. Job strain as a risk factor for clinical depression: Systematic review and meta-analysis with additional individual participant data. *Psychol Med* 47(8):1342–1356. DOI: 10.1017/S003329171600355X.

Milczarek, M., X. Irastorza, S. Leka et al. 2012. *Drivers and barriers for psychosocial risk management: An analysis of the findings of the European Survey of Enterprises on New and Emerging Risks (ESENER).* Luxembourg: Publications Office of the European Union.

Serra, C., M. C. Rodriguez, G. L. Delclos, M. Plana, L. I. Gómez López, and F. G. Benavides. 2007. Criteria and methods used for the assessment of fitness for work: A systematic review. *Occup Environ Med* 64(5):304–312. DOI: 10.1136/oem.2006.029397.

Tengland, P.-A. 2011. The concept of work ability. *J Occup Rehabil* 21(2):275–285. DOI: 10.1007/s10926-010-9269-x.

Tuomi, K., J. Ilmarinen, A. Jahkola, L. Katajarinne, and A. Tulkki. 1998. *Work ability index.* 2nd revised ed. Helsinki: Finnish Institute of Occupational Health.

2 Work Ability Index as a Tool of Assessment of the Possibilities to Perform Work

Teresa Makowiec-Dąbrowska
Nofer Institute of Occupational Medicine

Joanna Bugajska
Central Institute for Labour Protection –
National Research Institute

CONTENTS

ABOUT WORK ABILITY

Work ability is a notion comprising generally all components of psychophysical and mental capacity of a human being, necessary to perform a specific work. The degree of dependency between the ability to work and job demands is a decisive factor, predicting the real load related to work performance, the perception of stress and the development of work-related symptoms and conditions.

As mentioned in Chapter 1, the Finnish concept is one of the work ability concepts referring to subjective assessment of work ability and mental comfort, and objective health indicators. Work on this concept has contributed to the development of work ability assessment tool (*Work Ability Index* or *WAI*), enabling work ability rating in the following categories: excellent, good, moderate and poor (Tuomi et al. 1998).

WAI is calculated based on seven questions, five of which concern subjective assessment of current work ability compared with lifetime best (top form), physical and mental effort capacity required in the respondent's current job (work ability in relation to the demands of the job), estimated work impairment due to diseases, own prognosis of work ability 2 years from now and the so-called mental resources. The two remaining values are objective indicators of health status corresponding to the number of diseases diagnosed by a physician and the number of days of sick leave during the past year (12 months). The rating of answers is unequal. The highest scores (maximal score = 10) can be obtained for the current work ability rating as compared with lifetime best and the overall rating of meeting the job demands for physical and mental effort (work ability in relation to the demands of the job). The components of this element (with maximal score of 5 points) are multiplied by the coefficient depending on the respondent's exposure to workload, either physical or mental. In the case of physical workload only, the numerical value corresponding to meeting the demands related to physical effort is multiplied by 1.5, whereas the numerical values of rating related to meeting the mental demands are multiplied by 0.5. When mental workload is dominant, the values are measured the other way around. In the case of mixed workload, the values obtained from the rating of both components remain unchanged. The third and fourth most important elements of work ability (with maximal score of 7 points) are the number of diseases diagnosed by a physician and the worker's own prognosis of work ability 2 years from now. The maximal score obtained for the degree of impairment due to diseases is 6 points, whereas the maximal scores which can be obtained for sick leave during the last 12 months and the worker's mental resources are 5 and 1, 4 or 7, respectively. In each case, a higher rating indicates a better ability to work.

WAI is calculated based on the overall score obtained for these seven items. The range of index values is 7–49. The cut-off points for work ability category corresponding to the 15th percentile, the median value and the 85th percentile of index value distribution in the studied population of municipal workers aged 45–58 years are as follows: poor (7–27), moderate (28–36), good (37–43) and excellent (44–49) (Tuomi et al. 1998).

The rating of work ability based on WAI values is widely accepted. However, since work ability is highly dependent on age, this particular factor should be considered during selection of reference values. Therefore, it is suggested to increase the values obtained for younger workers before 30 years of age since their work ability is usually better (Kujala et al. 2005). Based on the analysis of the data reflecting work ability among 31-year-old employed men (n = 2021) and women (n = 1704) (about 7% of all 31-year-old employed Finns in 1997) using the same criteria of cut-off point selection for WAI categories, it was found that the 15th percentile was 37 points, the median value was 41 points and 85th percentile was 45 points. Kujala has proposed shift of the criteria for work ability category for younger workers to: 7–37 for poor work ability, 38–41 points for moderate work ability, 42–45 points for good work ability and 46–49 points for excellent work ability (Kujala et al. 2005).

ANALYSIS OF WORK ABILITY INDEX AND ITS COMPONENTS

WAI REPEATABILITY

WAI is characterized by a good short-term stability (repeatability). De Zwart et al. (2002) have shown that during the two studies conducted with a 4-week interval between, among a sample of 97 construction workers, no significant difference was noted in the mean value of WAI (40.4 compared with 39.9). The 66% coherence of respondent rating with one of the four WAI categories in both measurements was also satisfactory. The reproducibility of WAI values, measured with a 4-week interval between, was also studied in Thailand among a sample of 56 workers and rated as 0.71 (Kaewboonchoo and Ratanasiripong 2015).

The mean value of WAI reproducibility obtained by Yang et al. (2013) from the evaluation of their own research conducted in Korea was 0.70. The authors also found that the highest reproducibility values were obtained for WAI 1 corresponding to self-rating of the respondents' current work ability compared with lifetime best, WAI 2 corresponding to work ability in relation to the demands of the job and WAI 8 corresponding to self-rating of the respondent's mental resources. The remaining subscales were characterized by a significantly lower reproducibility.

A good reproducibility of WAI results, reflected by the mean value of 0.7, was also noted by Adel et al. (2019). The study also found a good internal consistency of WAI. The Cronbach alpha value obtained for this sample was 0.78.

WAI Factor Analyses

Radkiewicz et al. (2005) analysed the results of WAI in a very large group of nurses including more than 38.000 participants (NEXT-Study). The mean Cronbach alpha value corresponding to WAI internal consistency in the entire sample was 0.72, but some differences were noted between groups of participants from different countries, ranging from 0.54 in the Slovakian group to 0.79 in the Finnish group. The value obtained for the Polish group was 0.70. The factor analysis confirmed the homogeneity of WAI only in two among ten groups of different country representatives participating in the study. In the remaining groups, two factors with very similar structures were found. The first one can be interpreted as a subjective factor, whereas the second one is believed to be an objective work ability component.

Furthermore, the discriminant power of each WAI subscale was found to be varied. It was the highest for the work ability compared with lifetime best and work ability in relation to the demands of the job and the lowest for sick leave during the past year. Hence, the authors have concluded that this subscale can be safely removed from WAI. Such a suggestion, however, was not made in any other paper on WAI.

The Cronbach alpha value obtained from the analysis of work ability carried out by Abdolalizadeh et al. (2012) among 236 hospital workers, using WAI questionnaire, was 0.79. The factor analysis allowed the authors to determine three factors related to (1) self-rating of work ability, (2) mental resources and (3) the estimated work impairment due to diseases.

In the research conducted in Thailand among 2.744 participants, the Cronbach alpha value obtained for WAI was 0.66 (Kaewboonchoo and Ratanasiripong 2015). Three factors were determined based on factor analysis. The first factor was related to self-rating of current work ability compared with the lifetime best and work ability in relation to the physical and mental demands of the job; the second one was related to self-rating of four elements, namely number of current diseases diagnosed by a physician, estimated work impairment due to disease, sick leave during the past year and own prognosis of work ability 2 years from now; the third factor referred to the rating of own prognosis of work ability 2 years from now and three questions assessing the worker's mental resources.

Very similar results of WAI factor analysis were reported by Martinez et al. (2009) who analysed the outcome of the research carried out among 475 Brazilian workers. The first factor was composed of the elements of mental resource rating, the second one involved work ability assessment in relation to the physical and mental demands of a job, and the third one comprised the subscales of the estimated work impairment due to disease and own prognosis of work ability 2 years from now.

Makowiec-Dąbrowska et al. (2008) studied work ability among 1205 persons doing different jobs. The Cronbach alpha calculated based on the values obtained in the male group was not satisfactory, amounting only to 0.672, whereas the corresponding value obtained in the female group was 0.638. This fact may be explained by the relatively low correlation between WAI and its three items: sick leave during the past year, the number of current diseases diagnosed by a physician and the mental resources. In the male group, the correlation coefficients obtained for these parameters were 0.41, 0.51 and 0.47, respectively, whereas in the female group, the corresponding values were 0.39, 0.51 and 0.31, respectively. The correlation coefficients obtained for other items ranged from 0.62 to 0.71 in the male group and from 0.54 to 0.76 in the female group. The low correlation coefficient values corresponding to assessment of sick leave during the past year may result from the fact that, at least in Poland, sick leave is not a reliable indicator of health status. Being on sick leave or not depends not only on the worker's health status but also on extra-medical factors, such as economic conditions (a patient on sick leave receives only 80% of his or her earnings), family problems (child care), relations at workplace and attitudes towards work (worker's function, possibility of replacement, fear for being fired).

Martus et al. (2010) summarized the results of several studies conducted in Germany using WAI. A total of 371 workers doing different jobs (teachers, clerks, kindergarten teachers and managerial staff) were assessed. The goal of the analysis was to find out whether WAI was a one- or multidimensional index and to determine the stability of results in various studied populations. It was necessary to eliminate the managerial staff from the psychometric analyses of WAI since the ratings of own prognosis of work ability 2 years from now and mental resources were identical in all the respondents belonging to this occupational group and all of them obtained the highest scores. The results of the analyses indicated that WAI was not one dimensional. The result (factor) obtained from the three subscales, namely current work ability compared with the lifetime best, work ability in relation to the demands of the job and rating of mental resources, is defined as "subjective rating of work ability and resources"; the subscales referring to the number of current diseases diagnosed by a physician and sick leave during the past year form a factor defined as "health and the related factors". The remaining two subscales, namely estimated work impairment due to diseases and own prognosis of work ability 2 years from now, were not assigned to any of the above-mentioned categories. The authors have concluded that the application of overall WAI only is inappropriate for work ability analysis in individual subjects or the entire populations of workers.

Freyer et al. (2009) also point to the bifactor structure of WAI in their analysis of values reflecting work ability level in nearly 4,000 workers. The authors have concluded that the obtained data are best described using a model with two correlated factors. In this model, the items WAI 1, WAI 2, WAI 6 and WAI 7 form the factor referring to subjective assessment of the current and future work ability as well as the worker's individual resources. The second factor composed of WAI 3, WAI 4 and WAI 5 factors is related to health status. This model is analogous to that obtained by Radkiewicz et al. (2005) who reviewed the studies on work ability among nursing personnel.

Work Ability Index and Work Ability Score

Assessment of work ability using WAI may be difficult, due to the time-consuming assessment of the seven elements and respondents' problems with understanding some notions (mental resources or work ability definition), but also with determining the number of diseases diagnosed by a physician. Such an example is provided by Roelen et al. (2014a), whose analyses combined a high rate of missing data/responses (17%) and the length and complexity of WAI. Hence, attempts have been made to rate work ability with a simpler tool, such as one or several items of WAI.

El Fassi et al. (2013) compared work ability assessment based on the seven-item WAI with the assessment based only on the first item of WAI, called "Work Ability Score" (WAS). The occupational health service worker (over 12,000 workers aged 40–60 years) database was used for the study. A robust correlation was found between WAI and WAS (with Spearman's rank correlation coefficient rs = 0.63). It was also found that the same factors characterizing study participants and their work contributed to both the increase or decrease in WAI and WAS values (except the value corresponding to firm size, the factor which was not subjected to multivariate WAS analysis). Given the above findings, the authors conclude that WAS can be a useful tool for work ability screening.

The usefulness of WAS application for the assessment of work ability in physical workers was studied by Gupta et al. (2014) based on the percentage of heart rate reserve (% HRR) monitoring during several working days. A moderate negative correlation was found between WAS and % HRR only in several male subjects; this finding indicates that further research on this issue is needed. Nevertheless, the authors conclude that the obtained results indicate that WAS can also be a useful measure of physical load.

More sceptical attitudes to WAS assessment were presented by Roelen et al. (2014b) who used this index as a predictor of a future disability pension (DP). Their longitudinal study was performed within 2.3 years among a sample of over 11,000 construction workers including 3% of workers receiving a DP during the period of observation. Despite the fact that the mean values of both WAS and WAI were high and thus decreased the risk of DP, the WAI was found to better discriminate workers at high and low risk of DP. Therefore, the authors suggest using WAI for DP risk assessment in occupational health practice. However, the answer to a sole question concerning current work ability compared with lifetime best is much simpler for the respondent than completing the entire WAI questionnaire. Furthermore, WAS is easier to interpret as compared to overall WAI. A large-scale research using WAS may turn out to generate lower costs. This is the reason why the authors suggest using WAS as a main tool for work ability assessment and using the full-size WAI only in the respondents with low WAS scores.

A different opinion on using WAS as an alternative for WAI in DP risk assessment is presented by Jääskeläinen et al. (2016). The analysis of the results obtained from their research, carried out among municipal workers between the years 1981 and 2009, indicates a twofold or fivefold higher risk of DP in the subjects with moderate or poor work ability measured with WAI at the beginning of the study, as compared with those whose early WAI values reveal good or excellent work ability. This finding is also confirmed by the values corresponding to moderate or poor work ability according to WAS (6–7 or 0–5) in subjects whose DP risk has turned out to be 1.8 or 3.4 lower as compared to those with good or excellent work ability measured with WAS (8–10). At the same time, WAI was characterized by a slightly higher sensitivity, yet a slightly lower specificity than WAS. The authors conclude that WAS can be used as an alternative tool for WAI in DP prognosis.

The authors of the analysis predicting a long-term sickness absence have found that overall WAI is the best tool (Lundin et al. 2017) used for this purpose. Among the WAI components, there are three factors with proven predictive values related to long-term sick leave. These include *estimated work impairment due to diseases, number of current diseases diagnosed by a physician* and *current work ability compared with life time best* (WAS). Therefore, the authors suggest that these three items may be substitutes for WAI in population-based studies.

Ebener and Hasselhorn (2019) conducted an analysis to assess the usefulness of two items of WAI, namely WAI 1 (subjective assessment of current work ability compared with lifetime best) and WAI 2 (work ability assessment in relation to the physical and mental demands of the job). The data on work ability obtained for the female nursing staff participating in European NEXT-Study and nursing home workers in German 3Q-Study project were used for the analysis. The predictive value of WAI 1 and WAI 2 was assessed as compared with the self-rating of general health, occupational burnout and work exit decisions. In all cases, WAI 1 and WAI 2 values were moderately and similarly correlated with the above-mentioned variables which may have depended on work ability. Moreover, moderate to strong correlations were found between WAI 1 and WAI 2 values and overall WAI. Based on these findings, the authors suggest that WAI 1 and WAI 2 can replace overall WAI.

The aforementioned studies have always indicated a robust correlation between WAS and WAI. However, as suggested by Jääskeläinen et al. (2016), this correlation cannot be too high because WAS refers the current work ability levels to past values. If the subject's work ability has never been high and his/her current work ability level is similar, the WAS value of this parameter will be high, but it will not indicate a high level of work ability. Overall WAI, in turn, also contains other aspects of the current and future work ability; hence, it is more reliable than its single items.

APPLICATION OF WAI

Since WAI was developed, a large number of studies have been carried out using this tool (from 1991, there are 388 records including "Work Ability Index" in PubMed). The researchers dealing with work ability assessment using WAI have attempted to name all the factors and circumstances determining work ability levels and, next, to determine whether it is a predictor of future workers' occupational activity and health status (Ilmarinen 2009). The analyses of WAI determinants have been most often aimed at finding the factors responsible for work ability impairment and its improvement, namely the starting point for all intervention measures. Such analyses were carried out in many countries. They considered jobs, specific workloads, demographic characteristics and lifestyles of the studied workers. The systematic review of epidemiological studies, focused on factors determining work ability, has been presented by van den Berg et al. (2009). The authors conclude that, despite the differences in definitions of loading factors and the applied measurement techniques, the determining effect of high demands related to mental work, low level of self-control of situations at work and high requirements concerning physical effort was visible. Among the individual factors, the most important ones included the effect of age, lack of physical activity during leisure time, poor musculoskeletal capacity and obesity.

The predictive value of WAI was studied in relation to long-term sickness absence (e.g. Reeuwijk et al. 2015; Schouten et al. 2015), premature work exit (e.g. Salonen et al. 2003; Roelen et al. 2014b), disability retirement (e.g. Alavinia et al. 2009; Bethge et al. 2018) or need for rehabilitation (e.g. Bethge et al. 2012), as well as disability and/or mortality (e.g. Von Bonsdor et al. 2011).

REFERENCES

Abdolalizadeh, M., A. Arastoo, R. Ghsemzadeh, A. Montazeri, K. Ahmandi, and A. Azizi. 2012. The psychometric properties of an Iranian translation of the Work Ability Index (WAI) questionnaire. *J Occup Rehabil* 22(3):401–408. DOI: 10.1007/s10926-012-9355-3.

Adel, M., R. Akbar, and G. Ehsan. 2019. Validity and reliability of Work Ability Index (WAI) questionnaire among Iranian workers; a study in petrochemical and car manufacturing industries. *J Occup Health* 61(2):165–174. DOI: 10.1002/1348-9585.12028.

Alavinia, S. M., A. G. de Boer, J. C. van Duivenbooden, M. H. Frings-Dresen, and A. Burdorf. 2009. Determinants of work ability and its predictive value for disability. *Occup Med (Lond)* 59(1):32–37.

Bethge, M., F. M. Radoschewski, and C. Gutenbrunner. 2012. The Work Ability Index as a screening tool to identify the need for rehabilitation: Longitudinal findings from the Second German Sociomedical Panel of Employees. *J Rehabil Med* 44(11):980–987. DOI: 10.2340/16501977-1063.

Bethge, M., K. Spanier, E. Peters, E. Michel, and M. Radoschewski. 2018. Self-reported work ability predicts rehabilitation measures, disability pensions, other welfare benefits, and work participation: Longitudinal findings from a sample of German employees. *J Occup Rehabil* 28(3):495–503. DOI: 10.1007/s10926-017-9733-y.

de Zwart, B. C., M. H. Frings-Dresen, and J. C. van Duivenbooden. 2002. Test-retest reliability of the Work Ability Index questionnaire. *Occup Med (Lond)* 52(4):177–181.

Ebener, M., and H. M. Hasselhorn. 2019. Validation of short measures of work ability for research and employee surveys. *Int J Environ Res Public Health* 16(18):e3386. DOI: 10.3390/ijerph16183386.

El Fassi, M., V. Bocquet, N. Majery, M. L. Lair, S. Couffignal, and P. Mairiaux. 2013. Work ability assessment in a worker population: Comparison and determinants of Work Ability Index and work ability score. *BMC Public Health* 13:305. DOI: 10.1186/1471-2458-13-305.

Freyer, M., M. Formazin, and U. Rose. 2009. Factorial validity of the Work Ability Index among employees in Germany. *J Occup Rehabil* 29(2):433–442. DOI: 10.1007/s10926-018-9803-9.

Gupta, N., B. S. Jensen, K. Søgaard et al. 2014. Face validity of the single work ability item: Comparison with objectively measured heart rate reserve over several days. *Int J Environ Res Public Health* 11(5):5333–5348. DOI: 10.3390/ijerph110505333.

Ilmarinen, J. 2009. Work ability – A comprehensive concept for occupational health research and prevention. *Scand J Work Environ Health* 35(1):1–5. DOI:10.5271/sjweh.1304.

Jääskeläinen, A., J. Kausto, J. Seitsamo et al. 2016. Work Ability Index and perceived work ability as predictors of disability pension: A prospective study among Finnish municipal employees. *Scand J Work Environ Health* 42(6):490–499. DOI: 10.5271/sjweh.3598.

Kaewboonchoo, O., and P. Ratanasiripong. 2015. Psychometric properties of the Thai version of the Work Ability Index (Thai WAI). *J Occup Health* 57(4):371–377. DOI: 10.1539/joh.14-0173-OA.

Kujala, V., J. Remes, E. Ek, T. Tammelin, and J. Laitinen. 2005. Classification of Work Ability Index among young employees. *Occup Med (Lond)* 55(5):399–401. DOI:10.1093/occmed/kqi075.

Lundin, A., O. Leijon, M. Vaez, M. Hallgren, and M. Torgén. 2017. Predictive validity of the Work Ability Index and its individual items in the general population. *Scand J Public Health* 45(4):350–356. DOI: 10.1177/1403494817702759.

Makowiec-Dąbrowska, T., W. Koszada-Włodarczyk, A. Bortkiewicz et al. 2008. Zawodowe i pozazawodowe determinanty zdolności do pracy. *Med Pr* 59(1):9–24.

Martinez, M. C., R. Latorre Mdo, and F. M. Fischer. 2009. Validity and reliability of the Brazilian version of the Work Ability Index questionnaire. *Rev Saude Publica* 43(3):525–532.

Martus, P., O. Jakob, U. Rose, R. Seibt, and G. Freude. 2010. A comparative analysis of the Work Ability Index. *Occup Med (Lond)* 60(7):517–524. DOI: 10.1093/occmed/kqq093.

Radkiewicz, P., M. Widerszal-Bazyl, and The NEXT-Study Group. 2005. Psychometric properties of Work Ability Index in the light of comparative survey study. *Int Congr Ser* 1280:304–309.

Reeuwijk, K. G., S. J. Robroek, M. A. Niessen, R. A. Kraaijenhagen, Y. Vergouwe, and A. Burdorf. 2015. The prognostic value of the Work Ability Index for sickness absence among office workers. *PLoS One* 10(5):e0126969. DOI: 10.1371/journal.pone.0126969.

Roelen, C. A., M. W. Heymans, J. W. Twisk, J. J. van der Klink, J. W. Groothoff, and W. van Rhenen. 2014b. Work Ability Index as tool to identify workers at risk of premature work exit. *J Occup Rehabil* 24(4):747–754. DOI: 10.1007/s10926-014-9505-x.

Roelen, C. A., W. van Rhenen, J. W. Groothoff, J. J. van der Klink, J. W. Twisk, and M. W. Heymans. 2014a. Work ability as prognostic risk marker of disability pension: Single-item workability score versus multi-item Work Ability Index. *Scand J Work Environ Health* 40(4):428–431. DOI: 10.5271/sjweh.3428.

Salonen, P., H. Arola, C. H. Nygård, H. Huhtala, and A. M. Koivisto. 2003. Factors associated with premature departure from working life among ageing food industry employees. *Occup Med (Lond)* 53(1):65–68.

Schouten, L. S., C. I. Joling, J. W. van der Gulden, M. W. Heymans, U. Bültmann, and C. A. Roelen. 2015. Screening manual and office workers for risk of long-term sickness absence: Cut-off points for the Work Ability Index. *Scand J Work Environ Health* 41(1):36–42. DOI: 10.5271/sjweh.3465.

Tuomi, K., J. Ilmarinen, A. Jahkola, L. Katajarinne, and A. Tulkki. 1998. *Work Ability Index*. 2nd ed. Helsinki: Finnish Institute of Occupational Health.

van den Berg, T., L. Elders, B. de Zwart, and A. Burdorf. 2009. The effects of work-related and individual factors on the Work Ability Index: A systematic review. *Occup Environ Med* 66(4):211–220.

Von Bonsdor, M. B., J. Seitsamo, J. Ilmarinen, C.-H. Nygård, M. E. von Bonsdor, and T. Rantanen. 2011. Work ability in midlife as a predictor of mortality and disability in later life: A 28-year prospective follow-up study. *CMAJ* 183(4):E235–E242. DOI: 10.1503/cmaj.100713.

Yang, D. J., D. Kang, D. Kim et al. 2013. Reliability of self-administered Work Ability Index questionnaire among Korean workers. *Ergonomics* 56(11):1652–1657. DOI: 10.1080/00140139.2013.835073.

3 Dynamics of Changes in Work Ability According to the Type of Work, Age and Gender of Employees – Results of Research

Joanna Bugajska, Łukasz Baka, and Łukasz Kapica
Central Institute for Labour Protection –
National Research Institute

CONTENTS

INTRODUCTION

Work ability is the basis of a widely understood concept of physical and mental health. Presently, it is not so much recognized in terms of the lack of health-related contraindications as in terms of assessment of one's physical potential and psycho-motor and sensory capacities, allowing adequate work performance by an individual without risking his/her and other people's health and safety (Palmer and Cox 2007). One of the best-known concepts of work ability has been developed by a group of researchers from the Finnish Institute of Occupational Health (FIOH) who defined it as a balance between the worker's resources (including health status, functional abilities and skills) and the demands of the job (Ilmarinen 1991). A wider environmental context including family relations, society and legal regulations is also considered in work ability assessment. Based on the above concept, Work Ability Index (WAI) composed of seven subjectively assessed aspects has been developed. Presently, WAI is a widely used tool in work ability assessment to monitor the actions undertaken in enterprises and aimed at health and work ability promotions. WAI is also used to identify the needs related to prevention of a long-term sick leave, premature work exit and disability (Nygard et al. 2005; Kujala et al. 2006; von Bonsdorff et al. 2011; Schouten et al. 2015; Jaaskelainen et al. 2016).

Both Polish and foreign studies indicate that the important work ability determinants include worker's age and gender as well as work demand (Ilmarinen et al. 1997; Bugajska and Łastowiecka 2005; Malińska 2007; Makowiec-Dąbrowska et al. 2008). It means that the preventive programs and all the actions undertaken to maintain work ability of persons at any age should consider these factors.

The aim of the presented research was to assess the dynamics of changes in work ability in women and men from different age groups, doing work with different demands: physically demanding work (physical work), mentally demanding work (mental work) and physically and mentally demanding work (mixed work). The main WAI (WAI overall) and its two components (WAI 1: *current work ability compared with the lifetime best* and WAI 2: *work ability in relation to the demands of the job*) whose diagnostic values, comparable to WAI, had been confirmed in numerous occupational health research and employee surveys were accepted for the analysis (Alavinia et al. 2009; El Fassi et al. 2013; Roelen et al. 2014; Jaaskelainen et al. 2016; Lundin et al. 2017; Ebener and Hasselhorn 2019).

METHODS

PARTICIPANTS

The sample (N = 16.351) comprised men and women belonging to four age groups (18–30 years, 31–40 years, 41–50 years and 51–65 years), doing work with physical, mental and mixed (physical and mental) demands. The data were obtained from the database developed specifically for the assessment, including the data obtained from the research carried out at the Central Institute for Labour Protection – National Research Institute (CIOP-PIB). The group of physical workers included machine operators, production workers, cleaning staff, suppliers, warehouse workers, electricians, fitters and bakers, among others. The group of mental workers included teachers, librarians, physicians, clerks, IT professionals, specialists, logisticians and accountants, among others. The group of participants doing work with mixed demands included police officers, nurses, salespersons, sales representatives and drivers. The sample included 69% of men (n = 11,284) and 31% of women (n = 4832). As regards age distribution, the sample was divided into different age groups including 21.1% of 18–30-year-olds, 37.6% of 31–40-year-olds, 18.9% of 41–50-year-olds and 20.7% of 51–65-year-olds. Data shortages accounted for 1.8%. The participants performed three types of work with different demands namely: physical (13%), mental (24.4%) and mixed work (59.1%) demands. In approximately 3.6% of cases, data deficiencies were noted.

TOOL OF MEASUREMENT

Work ability was measured using WAI (Tuomi et al. 1998) questionnaire, translated into Polish (Pokorski 1998). The tool is described in Chapter 2. WAI overall and its two items, namely, WAI 1 – *current work ability compared with the lifetime best* and WAI 2 – *work ability in relation to the demands of the job*, were subjected to analysis.

ANALYTICAL PROCEDURE

The first stage of the study involved descriptive data analysis. Next, a multifactor variance analysis was carried out for the comparison of the results obtained from WAI and its two single items in three groups of participants doing work with physical, mental and physical and mental (mixed) demands, according to gender, age and gender–age interaction. When F-statistic for the effect of age, gender and interaction between these two variables was significant, a pairwise comparison test was carried out. When F-statistic for the effect of one of these variables (or age–gender interaction) was statistically insignificant, the Bonferroni post-hoc test was applied separately for statistically significant variables.

RESULTS

WAI OVERALL – RESULTS OF THE MULTIFACTOR ANALYSIS OF VARIANCE

The analysis of study results revealed a certain general trend of age-related decrease in mean WAI overall values in the studied occupational groups. The group of women doing work with mixed demands was the only exception (Figure 3.1).

The multifactor variance analysis showed the differences in WAI overall values obtained in the group of physical workers, by gender ($F = 33.82$; $p < 0.001$), age ($F = 37.99$; $p < 0.001$) and gender–age interaction ($F = 8.07$; $p < 0.001$). Similar regularities were noted in the group doing work with mixed demands, by age ($F = 77.42$; $p < 0.001$), gender ($F = 5.12$; $p < 0.05$) and age–gender interaction ($F = 6.60$; $p < 0.001$) which were discriminative factors for WAI overall. Slightly different values were obtained in the group of mental workers. In this group, only age was a discriminative factor for WAI overall ($F = 48.61$; $p < 0.001$). The gender-related differences, however, were statistically insignificant.

The analysis of the differences between the mean WAI overall values obtained in the studied age groups showed that their dynamics was different in the male and

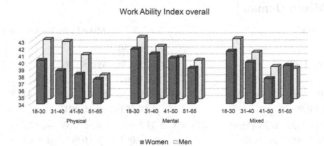

FIGURE 3.1 Mean WAI overall values obtained in workers belonging to three groups with different work demands, according to age and gender.

female physical workers (Table 3.1). In the male population, no differences were found only between the youngest group (18–30 years) and the group of males aged 31–40 years. Conversely, in the female population, significant differences were found only between the youngest (19–30 years) and the oldest (51–65 years) group.

The differences between the mean values of WAI overall in the group of workers performing mixed work indicate that their dynamics varies depending on gender (Table 3.2). In the male participants, no differences were noted only between the oldest group and the group of 41–50-year-old participants. Conversely, in the female participants, the lack of differences was noted only between the oldest group (51–65 years) and the workers aged 31–40 years.

The analysis of variance (ANOVA) showed that in the group of mental workers, only age (not gender) was a discriminant factor for WAI overall. Statistically significant differences were found between all age groups and a decrease in WAI overall values was noted with increasing age (Table 3.3).

TABLE 3.1
Results of Pairwise Comparison for WAI Overall in the Group of Physical Workers

Physical Work – Men

Age (years) 18–30			
0.29 ± 0.69	Age (years) 31–40		
$p = 0.90$			
2.17 ± 0.70	1.88 ± 0.61	Age (years) 41–50	
$p < 0.05$	**$p < 0.05$**		
5.11 ± 0.63	4.82 ± 0.53	2.94 ± 0.53	Age (years) 51–65
$p < 0.001$	**$p < 0.001$**	**$p < 0.001$**	

Physical Work – Women

Age (years) 18–30			
1.48 ± 0.74	Age (years) 31–40		
$p = 0.27$			
2.01 ± 0.81	0.53 ± 0.82	Age (years) 41–50	
$p = 0.81$	$p = 0.90$		
2.72 ± 0.58	1.25 ± 0.60	0.71 ± 0.69	Age (years) 51–65
$p < 0.001$	$p = 0.24$	$p = 0.90$	

Bold means statistically significant value ($p < 0.05$).

TABLE 3.2
Results of Pairwise Comparison of WAI Overall Values in the Group Doing Work with Mixed Demand

Mixed Work – Men

Age 18–30			
1.93 ± 0.21	Age 31–40		
$p < 0.001$			
3.97 ± 0.25	2.04 ± 0.21	Age 41–50	
$p < 0.001$	**$p < 0.001$**		
4.23 ± 0.34	2.30 ± 0.32	0.26 ± 0.34	Age 51–65
$p < 0.001$	**$p < 0.001$**	$p = 0.90$	

Mixed Work – Women

Age 18–30			
1.60 ± 0.38	Age 31–40		
$p < 0.001$			
3.95 ± 0.54	2.35 ± 0.50	Age 41–50	
$p < 0.001$	**$p < 0.001$**		
2.03 ± 0.42	0.43 ± 0.39	-1.92 ± 0.53	Age 51–65
$p < 0.001$	$p = 0.90$	**$p < 0.001$**	

Bold means statistically significant value ($p < 0.05$).

TABLE 3.3

Comparison of the Results of Bonferroni Post-hoc Test Obtained for WAI Overall in the Group of Mental Workers, by Age

		Mental Work		
Age 18–30				
$0.97 \pm 0.28\ \mathbf{p < 0.001}$	Age 31–40			
$2.10 \pm 0.32\ \mathbf{p < 0.001}$	$1.14 \pm 0.28\ \mathbf{p < 0.001}$	Age 41–50		
$2.98 \pm 0.27\ \mathbf{p < 0.001}$	$2.01 \pm 0.23\ \mathbf{p < 0.001}$	$0.87 \pm 0.28\ \mathbf{p < 0.05}$	Age 51–65	

Bold means statistically significant value (p < 0.05).

WAI 1 – *Current Work Ability Compared with the Lifetime Best* – Multifactor Variance Analysis

The results indicate that WAI values decrease with age; however, this regularity is mainly observed in physical workers and more seldom in those doing work with mental and mixed demands (Figure 3.2).

The analysis showed that in the group of physical workers, both age (F = 20.39; p<0.001) and gender (F = 18.84; p<0.001) were the discriminating factors for WAI 1. However, the interaction between these factors turned out insignificant, indicating a similar dynamics of changes in WAI 1 in the female and male participants. The Bonferroni post-hoc test revealed significantly lower WAI 1 values in women as compared to those obtained in men (M = 7.53 < M = 7.89). The differences in WAI 1 between the groups of participants doing work with physical, mental and mixed demands are presented in Table 3.4. In the groups doing work with mental and mixed demands, only age turned out to be a discriminating factor for WAI 1, which was reflected by the values – F = 5.08; p<0.05 and F 30.36; p<0.001 – obtained from these groups, respectively. No gender-related between-group differences were noted in these groups of workers.

Current work ability compared with the lifetime best

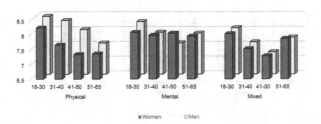

FIGURE 3.2 The mean values of work ability compared with the lifetime best obtained in three occupational groups of different age and gender.

TABLE 3.4

The Results of Bonferroni Post-hoc Test Obtained for WAI 1 (*Current Work Ability Compared with the Lifetime Best*), by Age in the Group of Workers Doing Work with Physical, Mental and Mixed Demands

Physical Work

Age 18–30			
0.29 ± 0.15 p = 0.29	Age 31–40		
0.51 ± 0.16 **p < 0.001**	0.21 ± 0.15 p = 0.90	Age 41–50	
0.88 ± 0.12 **p < 0.001**	0.58 ± 0.12 **p < 0.001**	0.37 ± 0.13 **p < 0.05**	Age 51–65

Mental Work

Age 18–30			
0.24 ± 0.09 **p < 0.05**	Age 31–40		
0.45 ± 0.10 **p < 0.001**	0.22 ± 0.09 p = 0.10	Age 41–50	
0.26 ± 0.08 **p < 0.05**	0.02 ± 0.07 p = 0.90	-0.19 ± 0.09 p = 0.18	Age 51–65

Mixed Work

Age 18–30			
0.48 ± 0.06 **p < 0.001**	Age 31–40		
0.80 ± 0.07 **p < 0.001**	0.33 ± 0.08 **p < 0.05**	Age 41–50	
0.25 ± 0.08 **p < 0.05**	-0.24 ± 0.07 **p < 0.001**	-0.56 ± 0.08 **p < 0.001**	Age 51–65

Bold means statistically significant value (p < 0.05).

WAI 2 – WORK ABILITY IN RELATION TO THE DEMANDS OF THE JOB – RESULTS OF MULTIFACTOR VARIANCE ANALYSIS

Figure 3.3 presents the results obtained for work ability in relations to the demands of the job. As in the case of WAI 1 analysis presented above, in both the male and female participants, a decrease in WAI 2 is noted; this regularity, however, concerns only physical work and, to a lower extent, mental and mixed work demands.

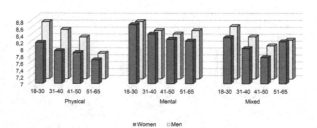

FIGURE 3.3 Mean WAI 2 work ability in relation to the demands of the job for workers performing three types of work, by age and gender.

TABLE 3.5

WAI 2 Results of Bonferroni Post-hoc Test Results in Relation to the Demands of the Job in Participants Performing Physical, Mental and Mixed Work, by Age

	Physical Work		
Age 18–30			
0.17 ± 0.13 p = −0.90	Age 31–40		
0.27 ± 0.14 p = 0.29	0.10 ± 0.13 p = 0.90	Age 41–50	
0.68 ± 0.11 **p < 0.001**	0.51 ± 0.10 **p < 0.001**	0.41 ± 0.11 **p < 0.001**	Age 51–65
	Mental Work		
Age 18–30			
0.27 ± 0.07 **p < 0.001**	Age 31–40		
0.39 ± 0.08 **p < 0.001**	0.12 ± 0.07 p < 0.54	Age 41–50	
0.36 ± 0.07 **p < 0.001**	0.09 ± 0.06 p < 0.66	−0.03 ± 0.07 p < 0.90	Age 51–65
	Mixed Work		
Age 18–30			
0.29 ± 0.04 **p < 0.001**	Age 31–40		
0.54 ± 0.04 **p < 0.001**	0.24 ± 0.04 **p < 0.001**	Age 41–50	
0.30 ± 0.06 **p < 0.001**	0.01 ± 0.05 p = 0.90	−0.24 ± 0.06 **p < 0.001**	Age 51–65

Bold means statistically significant value (p < 0.05).

The analysis showed that in the groups of physical workers, both age (F = 18.71; p < 0.001) and gender (F = 13.15; p < 0.001) were the discriminating factors for WAI 2; however, the interaction between these factors (age x gender) turned out insignificant. This finding indicates that the level of differences in WAI 2 between the age groups is similar in the studied male and female population. Similar regularities were noted in workers with mixed job demands; both age (F = 31.62; p < 0.001) and gender (F = 8.77; p < 0.05) were the independently discriminating factors for WAI 2, and the interaction between these factors was statistically insignificant as well. The Bonferroni post-hoc test results corresponding to age and gender indicated significantly lower WAI 2 values in women as compared with those obtained in men (M = 7.82 < M = 8.06 and M = 8.11 < M = 8.23 for women and men, respectively).

Conversely, among mental workers, only age turned out to be a discriminating factor for WAI 2 (F = 11.57; p < 0.001). The gender-related differences were statistically insignificant. Table 3.5 presents the age-related differences in work ability in relation to the demands of the job.

DISCUSSION

The aim of the research was to compare the dynamics of changes in WAI values in relation to age of the female and male participants doing work with physical, mental and mixed demands.

The analyses were carried out among a relatively large sample (N = 16.351) of Polish workers. The global WAI overall and its two items, namely, WAI 1 (*current work ability compared with the lifetime best*) and WAI 2 (*work ability in relation to the (physical/mental) demands of the job*), were selected for comparative analysis. The selection of these tools was based on literature review. According to Lundin et al. (2017), WAI 1 is one of the three items of WAI overall, considered to be a suitable tool, used in public health surveys. Ebener and Hasselhorn (2019), in turn, based on the results of their cross-sectional studies as well as 12-month prospective studies carried out among two occupational groups (nurses and non-nursing staff) confirm the possibility of replacing WAI overall with single WAI items, such as WAI 1, reflecting *current work ability compared with the lifetime best*, and WAI 2, reflecting *work ability in relation to the (physical/mental) demands of the job*, especially in occupational health research and worker surveys.

The research results showed the highest levels of work ability in workers with mental work demands, whereas the lowest corresponding values were noted in those with physical work demands. The abovementioned regularity was found in both WAI overall values and its two single items, WAI 1 and WAI 2.

Age turned out to be the main, statistically significant discriminating factor for overall work ability (WAI overall). However, in groups of workers with the three aforementioned job demands, the dynamics of WAI overall was found to be slightly different. Generally, the values corresponding to WAI overall, as well as WAI 1 and WAI 2, were found to decrease with increasing age. This trend was mainly noted in physical workers. A different general trend was observed among the participants doing work with mental demands. The highest WAI 1 values were noted in the youngest participants (aged 18–25 years), whereas the second highest corresponding values were obtained in the oldest workers. The highest values WAI 2 were noted in the youngest workers. No differences in this index values were observed between the oldest workers and those aged 31–40 and 41–50 years. It is difficult to find specific general trends in groups of workers with mixed work demands.

The results of the conducted analyses indicate that gender was a discriminating factor for WAI overall in workers with physical and mixed work demands. In these groups, slightly lower values of WAI overall and WAI 2 were noted in women. The values obtained by mental workers revealed no gender-related differences. In case of WAI 1, gender-related differences were noted only in physical workers. The values obtained in women belonging to this group revealed a slightly lower overall level of this variable. No significant differences in WAI 1 were found in women and men doing work with mental and mixed demands.

The results also showed that the effect of interaction between the variables corresponding to gender and age was only significant for the differences in WAI overall in workers with physical and mixed job demands. The results showed that WAI overall for both sexes decreased with age, however the trend is much more pronounced among men. Among workers with mental job demands, no statistically significant gender-related differences in this parameter were noted. The lack of a statistically significant effect of the interaction between gender and age variables, despite the statistically significant discriminating effect of age and gender (separately) on these indices, indicates a similar age-related decrease in WAI 1 and WAI 2 values in the male and female participants. The Finnish follow-up study, carried out among the population of workers over 45, has found good and excellent work ability levels in about 60% of the participants; the results also indicate a decrease in work ability level within 11 consecutive years in 30% of participants and an increase in this parameter in about 10% of the sample. Gender was not a discriminating factor for these changes. Furthermore, the changes were relatively similar in the studied occupational groups. Despite this fact, an improvement of work ability was more often observed in women performing physically demanding jobs and men performing mentally demanding jobs. On the contrary, a decrease in work ability was observed in men having mixed mentally and physically demanding jobs, particularly in male transport workers and female nursing staff (Tuomi 1997, Ilmarinen et al. 1997). These results indicate that there is no single trajectory of age-related changes in work ability of the working population. Work ability depends on multiple, both individual and work-related factors. However, some typical patterns of changes in work ability can be observed depending on the type of work performed and gender.

Our studies are cross-sectional, which is an obvious limitation in drawing conclusions, making it difficult to compare the results with the values obtained from earlier research. Nevertheless, based on our findings, we may conclude that there are clear differences in the dynamics of WAI overall between men and women belonging to various age groups with physical and mental work demands. No such differences were observed in the case of mental work demands. There are no statistically significant differences in WAI overall between the three age groups of women, with the youngest ones aged 31 (31–40; 41–50; 51–65), whereas in the male population, such differences can be found between each of the corresponding age groups. The lack of differences in the dynamics of changes in WAI 1 and WAI 2 between men and women belonging to different age groups with different work demands can be explained by the nature of these items of WAI. Both of these items selected for the analysis depend on subjective assessment and the worker's individual resources (Freyer et al. 2019). Additionally, in the case of WAI 2, the study result may indicate a very positive aspect. It may show that in both the studied men and women, the age-related changes result from a good adjustment of job demands to the worker's resources.

In summary, we can conclude that physical effort plays a more important role in the dynamics of age- and gender-related changes in work ability than mental effort. This study has revealed the lowest work ability levels and the highest age-related decrease in work ability in physical workers. Besides, in this group of workers, the biggest differences in work ability are noted between men and women. On contrary, in mental workers, the age-related changes in work ability are less pronounced, and there are no apparent differences in work ability levels between men and women. In case of workers with mixed work demand, we may assume that mental effort is responsible for the decreased effect of physical effort on work ability. In this group of workers, higher values of WAI, WAI 1 and WAI 2 are observed in the oldest participants as compared with their younger counterparts. This phenomenon may be due to the healthy worker effect or a better adjustment to the demand of the job.

We may also assume that the older workers belonging to the occupational groups with mixed work demands (policemen, nurses, dealers, sales representatives, drivers) have found jobs requiring less physical effort and therefore their subjective rating of work ability is higher.

REFERENCES

Alavinia, S. M., de Boer, A. G. E., van Duivenbooden, J. C., Frings-Dresen, M. H. W., and A. Burdorf. 2009. Determinants of work ability and its predictive value for disability. *Occup Med-Oxf* 59(1):32–37. DOI: 10.1093/occmed/kqn148.

Bugajska, J., and E. Łastowiecka. 2005. Life style, work environment factors and work ability in different occupations. *Int Congr Ser* 1280:247–252.

Ebener, M., and H. M. Hasselhorn. 2019. Validation of short measures of work ability for research and employee surveys. *Int J Environ Res Public Health* 16(18):e3386. DOI: 10.3390/ijerph16183386.

El Fassi, M., V. Bocquet, N. Majery et al. 2013. Work ability assessment in a worker population: Comparison and determinants of Work Ability Index and Work Ability score. *BMC Public Health* 13:e305. DOI: 10.1186/1471-2458-13-305.

Freyer, M., M. Formazin, and U. Rose. 2019. Factorial validity of the Work Ability Index among employees. *J Occup Rehabil* 29(2):433–442.

Ilmarinen, J. 1991. The aging worker. *Scand J Work Environ Health* 17(Suppl 1):1–141.

Ilmarinen, J., K. Tuomi, and M. Klockars. 1997. Changes in the work ability of active employees over an 11-year period. *Scand J Work Environ Health* 23(Suppl 1):49–57.

Jaaskelainen, A., J. Kausto, J. Seitsamo et al. 2016. Work Ability Index and perceived work ability as predictors of disability pension: A prospective study among Finnish municipal employees. *Scand J Work Environ Health* 42(6):490–499. DOI: 10.5271/sjweh.3598.

Kujala, V., T. Tammelin, J. Remes et al. 2006. Work Ability Index of young employees and their sickness absence during the following year. *Scand J Work Environ Health* 32(1):75–84.

Lundin, A., O. Leijon, M. Vaez, M. Hallgren, and M. Torgén. 2017. Predictive validity of the Work Ability Index and its individual items in the general population. *Scand J Public Health* 45(4):350–356. DOI: 10.1177/1403494817702759.

Makowiec-Dąbrowska, T., W. Koszada-Włodarczyk, A. Bortkiewicz et al. 2008. Zawodowe i pozazawodowe determinanty zdolności do pracy. *Med Pr* 59(1):9–24.

Malińska, M. 2007. Ocena zdolności do pracy pracowników starszych wg WAI – wyniki wybranych polskich badań [Work Ability Index (WAI) of older workers: A review of selected Polish research]. *Bezpieczeństwo Pracy – Nauka i Praktyka* 5:15–20.

Nygard, C. H., H. Arola, A. Siukola et al. 2005. Perceived work ability and certified sickness absence among workers in a food industry. *Int Congr Ser* 1280:296–300.

Palmer, K., and R. Cox. 2007. A general framework for assessing fitness for work. In *Fitness for Work. A Medical Aspects*, eds. K. Palmer, R. Cox, and I. Brown, 1–20. Oxford: Oxford University Press.

Pokorski, J. 1998. *Indeks Zdolności do Pracy (WAI)*. Kraków: Wydawnictwo Uniwersytetu Jagiellońskiego.

Roelen, C. A., W. van Rhenen, J. W. Groothoff, J. J. van der Klink, J. W. Twisk, and M. W. Heymans. 2014. Work ability as prognostic risk marker of disability pension: Single-item workability score versus multi-item Work Ability Index. *Scand J Work Environ Health* 40(4):428–431. DOI: 10.5271/sjweh.3428.

Schouten, L. S., C. I. Joling, J. W. J. van der Gulden et al. 2015. Screening manual and office workers for risk of long-term sickness absence: Cut-off points for the Work Ability Index. *Scand J Work Environ Health* 41(1):36–42.

Tuomi, K., ed. 1997. Eleven-year follow-up of aging workers. *Scand J Work Environ Health* 23(Suppl 1):1–71.

Tuomi, K., Ilmarinen, J., Jahkola, A., Katajarinne, L., and Tulkki, A. (1998). *Work Ability Index*. 2nd revised version. Helsinki: Finnish Institute of Occupational Health.

von Bonsdorff, M. B., J. Seitsamo, J. Ilmarinen et al. 2011. Work ability in midlife as a predictor of mortality and disability in later life: A 28-year prospective follow-up study. *Can Med Assoc J* 183(4):E235–E242.

4 Work Ability Index and Its Relationships to Factors Characterizing Work, Non-professional Loads and Individual Factors – Results of Research

Teresa Makowiec-Dąbrowska
Nofer Institute of Occupational Medicine

CONTENTS

INTRODUCTION

People can successfully do their jobs for many years without adverse health-related events when they are not exposed to excessive load. Workload results from the lack of balance between effort intensity, norms and other job requirements, the characteristics of chemical and physical parameters of working environment and psychosocial conditions, and the worker's individual resources. The latter depend on knowledge and skills (education level) as well as experience, physical and mental condition, lifestyle, health status and extraprofessional burdens. Age is an important factor modifying the amount of such resources in different directions and to a varying extent.

Objective measurement of workload considering the plethora of factors affecting its volume is virtually impossible. An assessment of workload by workers themselves seems to be a reasonable solution in this case. A worker can most reliably assess his or her work ability in relation to job requirements. Work Ability Index (WAI) is a very good tool for subjective assessment of man's work ability (Tuomi et al. 1994). It allows a quantitative assessment of subjective workload at workplace, the present and future physical and mental potential for overcoming the workload-related problems, as well as health status and its impact on work ability. WAI has good psychometric properties and a confirmed repeatability of results obtained from subsequent measurements (Radkiewicz and Widerszal-Bazyl 2005; De Zwart et al. 2002). It is positively correlated with well-being (Sjögren-Rönkä et al. 2002; Tavakoli-Fard et al. 2016) and is a predictor of prognosis for further ability to continue work (Krause et al. 1997). Therefore, it is extremely important to maintain or improve work ability to increase participation in job-related activities and extend the period of occupational activity in ageing population. The knowledge of factors affecting work ability levels makes it possible to plan rational actions aimed at meeting this goal and to prepare adequate intervention measures. Van den Berg et al. (2009), based on the review of 20 papers (describing 14 cross-sectional and 6 prospective studies), suggest that the factors responsible for low work ability levels measured using WAI included unfriendly working environment and excessive effort at work, high requirements relating to mental activities and the lack of autonomy, as well as the factors determining the worker's abilities, such as the lack of intense physical activity in free time, poor capacity of the musculoskeletal system, a more advanced age and obesity (van den Berg et al. 2009). The problems or hard living condition unrelated to work are also considered to be the factor responsible for decreased WAI values (Pohjonen 2001).

In Poland, the pilot study using WAI was carried out in 2000, (Makowiec-Dąbrowska et al. 2000), and an extensive study on subjective perception of work ability in persons of different age, doing different jobs and therefore exposed to a variety of factors was carried out as part of the task: "Occupational and extra-occupational work ability determinants in older workers", being, in turn, part of the research project commissioned by the Ministry of Family, Labor and Social Policy in Poland. The title of the project was "Occupational activity of workers in the aspect of ageing society problems". The obtained parameters were subjected to a multidirectional analysis and will be presented in this chapter as individual factors and/or the factors characterizing a given job and working conditions which turned out to be work ability predictors in men and women in the three work content groups: physical, mixed (mental and physical) and mental.

MATERIALS AND METHODS

It was planned to carry out the study in several workplaces (hypermarkets, post office, fire department, brown coal mine, metal manufactory, Municipal Transport Company, cosmetics factory) in a sample of 1800 workers. A self-rating questionnaire was the basic research tool used in this study. The participants were recruited by occupational safety and health services who distributed the questionnaires to all workers with minimum 2-year work experience in the current position.

Thousand seven hundred and four completed questionnaires were returned (94.6%), and 501 questionnaires containing inadequate information were eliminated. Finally, 1205 questionnaires were subjected to analysis.

The questionnaires used for the study consisted of:

- WAI questionnaire for workability (WA) assessment
- Subjective assessment of work-related stress
- CIS20-R questionnaire for prolonged fatigue assessment
- Questions allowing to obtain information about job-related factors, work-related fatigue, non-professional loads and lifestyle.

The Polish version of WAI questionnaire developed in the 90s at Finish Institute of Occupational Health (FIOH) (Tuomi et al. 1994) and applied in this study has been translated by Pokorski.

The questionnaire for subjective assessment of work-related stress was developed by Dudek et al. (1999). It is composed of 60 items related to job requirements and characteristics, assessed using the following scale:

1. The feature does not exist or is unrelated to my position.
2. The feature exists, but I don't mind it, I don't find it annoying.
3. Sometimes it's annoying or disturbing.
4. It's often annoying or disturbing.
5. It's annoying during the entire time at work, and sometimes, it annoys me when I am at home.

The score corresponds to perceived stress level; the theoretical range of scores is 60–300. The mean Cronbach α value for the entire questionnaire is 0.84 (Dudek et al. 1999).

Prolonged fatigue assessment was performed using CIS20R (Checklist Individual Strength) (Vercoulen et al. 1994). Its Polish version was developed by Makowiec-Dąbrowska and Koszada-Włodarczyk (2006). The questionnaire contains 20 statements presenting symptoms indicating fatigue. A participant has to assess in 1–7-point scale to what extent each of the statement corresponds to his/her perception (state) within the 2 weeks preceding the study. Fatigue index is calculated based on the scores obtained for each statement: the theoretical range of scores is 7–140. The value of Cronbach's α in Polish version of the whole questionnaire was 0.912.

Job characteristics comprised working time in the permanent workplace, working overtime and/or in another workplace, working system (one or two shifts, three shifts, night work and working time longer than 8 hours, also at night) and assessment of energy expenditure at work (based on the measurement of lung ventilation at work). According to Polish classification of energy expenditure in men, light work corresponds to energy expenditure up to 800 kcal within the 8-hour shift, medium-hard work corresponds to energy expenditure of 1500 kcal and hard work corresponds to energy expenditure above 1500 kcal. The corresponding reference values for women are 700 kcal, up to 1000 kcal and over 1000 kcal.

Next, the participants provided information on exposure (or no exposure) to the following factors at work: awkward posture (including the time of exposure, if so), bending, squatting, high reach, far reach, too high or too low working surface, generally uncomfortable work post, malfunctioning equipment, a job requiring only standing position, too little place at the work post, lifting or moving heavy loads of 13–20 kg and over 20 kg and other difficulties/arduousness (generally large amount of work, imposed pace of work, uneven pace of work, high repeatability of movements, generally heavy physical exertion, periodically heavy physical exertion, continuous walking, prolonged standing and/or sitting).

In the assessment of the working environment, the participants were asked whether they were exposed to chemical agents, mineral dust, high temperatures (resulting in sweating also while not working), cold (temperature below 10°C), humidity, inadequate lighting or noise, making it difficult to understand speech and whole body vibration. In order to assess work fatigue, the respondents were asked to indicate what proportion (in per cent) of the strongest fatigue they ever experienced after work on the day preceding the survey. The range of scores obtained from possible answers was 0 to 100. Moreover, the participants assessed the physical effort at work in relation to own possibilities using a 4-point scale (1 – I was able to work harder; 2 – effort at work fits me; 3 – sometimes, my job takes too much effort; 4 – my job always takes too much effort). Next, the respondents indicated whether due to fatigue after a working day they gave up their obligatory non-professional activities or hobbies. The following scores were obtained for each item: (1) I did not feel that tired; (2) yes, but very rarely; (3) yes, quite often; and (4) yes, very often.

The participants assessed their non-professional duties using a 5-point scale, selecting one suitable answer from the following: 1 – these duties are not burdensome for me – to 5 – in my opinion, the workload is very high, as well as their opportunities for rest from 1 to 4 points where (1) I can always rest long enough when I feel tired; (2) there are some limitations, but basically the amount of rest is sufficient; (3) I do not have the appropriate conditions for rest; and (4) I have neither appropriate conditions nor opportunities for rest and the average length of sleep on working days. They also rated the incidence of the following sleep disturbances:

- Difficulties in falling asleep
- Difficulties in waking up
- Waking up at night and difficulties in falling asleep again
- Nightmares
- Feeling sleepy after waking up
- Waking up too early
- The feeling of not getting enough rest after waking up
- Feeling sleepy at work and during leisure time
- Sensation that you have sand under eyelids after waking up
- Heavy head feeling after waking up
- Snoring during sleep.

The above list of sleep disturbances had been developed for this study. Each statement was scored on a scale from 1 (never, rarely) to 4 (very often), and the total score corresponded to sleep disturbance index. The theoretical scope of the index values is 11–44.

The data corresponding to lifestyle included the following:

1. The number of hours spent (doing):
   - High-intensity recreational exercise (e.g. sport training fitness club)
   - Moderate-intensity recreational exercise (e.g. riding a bike, aerobic, jogging)
   - Low-intensity recreational exercise (e.g. walking)
   - Passive rest in a sitting position (e.g. reading, watching TV, meeting with friends)
   - Resting in a lying position.
2. Tobacco smoking (being a daily smoker, smoking one or more cigarettes per day during the past 30 days at a minimum).
3. Caffeine consumption mg/day.
4. Frequency of getting drunk in 1 (never) to 5 (it happens several times a week).

In addition, the respondents reported their eating habits (preference for fatty or salty foods) and completed a table showing the main food ingredients (21 items) and their frequency of consumption (daily, several times a week, sometimes, never). The frequency of consumption was assessed in 1 (completely in disagreement with healthy nutrition rules) to 4 (totally in agreement with healthy nutrition rules) scale and the nutritional quality index value reflected the overall score obtained by each participant. This method was based on the *Food Frequency Questionnaire* (FFQ) which did not take into account the quantities of the consumed products (Cade et al. 2002).

STATISTICAL ANALYSIS

Descriptive statistics were applied to assess work ability in the studied sample as well as the job-related factors, non-professional load and individual characteristics of the sample. The mean values and standard deviations were calculated for continuous variables. The frequency of exposure to each load was expressed in percentage. For the comparison of the female and male groups, the analysis of variance and chi-squared test were applied, respectively.

The correlations between individual characteristics, work-related factors, extracurricular burden and lifestyle, and WAI and its components were assessed using multiple stepwise regression model. In this model, independent variables, which are not correlated with the dependent variable, are gradually subtracted until a subset of variables significantly correlated to the dependent variable is obtained. This method also allows to eliminate independent variables whose correlation with the dependent variable results only from a significant correlation between these variables. The results of such analysis are most often described as a set of the best predictors of the dependent variable.

All the analyses were carried out separately in male and female participants groups according to the dominant workload at workplace because numerous parameters revealed significant between-group differences.

RESULTS

SAMPLE CHARACTERISTICS

A total of 669 men aged 20–65 years and 536 women aged 18–63 years participated in the study. The majority of male participants were physical workers (this group was composed of over half of trade workers, the majority of cold storage workers, firemen, miners, metal manufactory workers and drivers). Fewer workers were exposed to mixed workload (trade workers, post office staff and firemen), whereas the fewest workers were exposed to mental load (middle-level managers). Among the studied females, the largest group included those exposed to mixed workload (most of the health service workers, trade workers, post office staff, some cold storage workers and cosmetics factory workers). There were fewer physical workers (some trade workers and post office staff workers, most of cold storage workers). The smallest group included women doing mental work (middle-level managers from the above-mentioned workplaces and bank clerks). Table 4.1 presents the detailed sample characteristics such as age, education level, seniority and work-related and extracurricular load indicators depending on the type of load.

Among both the male and the female, the oldest group with the longest seniority were people performed mental work; they also had the largest number of years of education.

In the male group, the least strenuous work (the lowest energy expenditure) was noted in mental workers, whereas the physical workers did the hardest work; although all the mean values obtained in this group were within the range of moderately hard work, the between-group differences were statistically significant. Similar dependencies on the type of work were found in the group of women, except for those performing work with the dominant physical load it was medium-heavy work, and for those performing work with mixed and mental load – light. For almost 40% of men exposed to physical or mixed workload and only 20% of men exposed to mental workload, physical effort was sometimes at least too high. Over 60% of women exposed to physical workload, about 50% of women exposed to mixed workload and about 30% women exposed to mental workload physical effort was too high, at least sometimes.

Physical workers (men and women) were more often than other workers exposed to adverse factors in their working environment. Their workstation was more often uncomfortable, and they more often assumed awkward postures at work.

The highest rating of occupational stress was found in men exposed to mental workload, and the lowest corresponding values were found in men exposed to physical workload. Conversely, among the female participants, the highest stress rating was found in those exposed to physical workload, whereas the lowest corresponding values were found in those exposed to mixed workload; however, the differences were statistically insignificant.

The highest fatigue levels after work were reported by men exposed to mixed workload, whereas those exposed to mental workload reported the lowest levels of fatigue. The kind of workload in the male group did not differentiate their prolonged fatigue levels. In the female group, the levels of fatigue after work and prolonged fatigue levels did not depend on their exposure to workload at workplace.

TABLE 4.1

Characteristics of the Studied Sample

| | Men | | | | Women | | | |
| | Work Content | | | | Work Content | | | |
	Mostly Physical N = 393	Mixed N = 230	Mostly Mental N = 46	p	Mostly Physical N = 124	Mixed N = 301	Mostly Mental N = 111	p
Age [years]	39.35 ± 10.87	38.94 ± 10.77	42.50 ± 7.96	0.1168	35.26 ± 8.99	37.13 ± 9.78	40.67 ± 9.88	0.0001
Number of years of education	11.41 ± 1.99	12.17 ± 1.74	13.55 ± 3.01	0.0000	11.96 ± 2.11	12.71 ± 1.68	13.46 ± 2.13	0.0000
General seniority [years]	20.24 ± 11.13	18.89 ± 11.69	22.38 ± 8.94	0.0586	14.35 ± 9.30	16.68 ± 9.83	19.88 ± 10.24	0.0001
Seniority in the workplace [years]	13.13 ± 10.19	9.74 ± 8.13	5.57 ± 8.91	0.0000	8.09 ± 8.27	9.75 ± 8.45	12.56 ± 9.92	0.0004
Seniority in the work post [years]	10.86 ± 9.39	6.38 ± 6.09	0.20 ± 8.98	0.0000	7.74 ± 8.19	8.98 ± 7.94	9.14 ± 7.54	0.2799
Work time [h]	7.97 ± 0.52	8.01 ± 0.84	8.13 ± 0.52	0.5690	7.92 ± 0.53	8.37 ± 1.73	7.93 ± 0.58	0.0009
Working overtime [% of persons]	35.37	36.96	25.65	0.0333	16.13	13.95	18.92	0.4541
Additional job [% of persons]	4.33	10.00	23.91	0.0000	2.42	2.66	4.40	0.5677
Night shift work [% of persons]	24.55	48.28	18.62	0.0000	20.6	22.92	2.73	0.0000
Energy expenditure [kcal/8h shift]	1326.4 ± 323.0	1094.6 ± 320.5	816.6 ± 184.7	0.0000	892.5 ± 235.9	673.1 ± 268.0	407.8 ± 96.0	0.0000
Frequent or regular exposure to chemical agents or physical factors [% of persons]	33.84	18.69	10.87	0.0008	36.29	14.28	5.24	0.0000
Uncomfortable work post [% persons]	64.35	15.61	15.38	0.0000	18.25	11.41	8.95	0.0000
Awkward posture 50% of work time or longer [% of persons]	24.09	6.76	24.11	0.2950	32.97	23.18	4.29	0.0002
Carrying loads 13–20 kg [% of persons]	34.09	41.84	0.00	0.0043	24.19	14.29	2.70	0.0001
Carrying loads >20 kg [% of persons]	35.45	20.57	6.67	0.0015	18.55	18.60	0.00	0.0001

(Continued)

TABLE 4.1 (Continued)
Characteristics of the Studied Sample

	Men				Women			
	Work Content				Work Content			
	Mostly Physical N = 393	Mixed N = 230	Mostly Mental N = 46	p	Mostly Physical N = 124	Mixed N = 301	Mostly Mental N = 111	p
Other difficulties/ arduousness [% of persons]	46.06	56.96	56.52	**0.0018**	61.29	61.80	46.85	**0.0202**
Occupational stress [score 60–300]	112.38 ± 29.81	109.35 ± 26.89	125.27 ± 28.33	**0.0194**	118.01 ± 30.83	115.05 ± 27.37	117.01 ± 26.31	0.5869
Assessment of the physical effort at work in relation to own possibilities [% of persons]:				0.2318				**0.0000**
1. I was able to work harder	7.31	13.10	10.00		4.03	4.01	10.91	
2. Effort at work fits me	53.42	51.19	66.67		33.06	48.18	60.91	
3. Sometimes, my job takes too much effort	36.06	34.52	20.00		55.65	44.82	28.18	
4. My job always takes too much effort	3.20	1.19	3.33		7.26	3.01	0.00	
Fatigue after work [as per cent of the highest fatigue perceived]	45.74 ± 22.84	47.99 ± 22.81	35.87 ± 21.14	**0.0049**	52.93 ± 23.92	52.96 ± 23.98	55.71 ± 23.41	0.5583
Giving up non-professional duties – often [% of persons]	21.10	19.53	13.33	0.9178	34.07	30.17	31.42	0.7078
Giving up hobbies – often [% of persons]	30.14	30.18	30.00	0.9676	50.55	41.38	50.00	0.7166
High load connected with household chores [% of persons]	17.18	28.82	33.33	**0.0003**	6.46	7.69	9.01	0.3560
Prolonged fatigue [score 7–140]	65.24 ± 25.34	66.84 ± 21.43	67.50 ± 20.85	0.7964	75.45 ± 25.04	71.06 ± 29.75	66.22 ± 26.26	0.1447

(Continued)

TABLE 4.1 (Continued)
Characteristics of the Studied Sample

	Men				Women			
	Work Content				Work Content			
	Mostly Physical N = 393	Mixed N = 230	Mostly Mental N = 46	p	Mostly Physical N = 124	Mixed N = 301	Mostly Mental N = 111	p
No possibility of resting at home [% of persons]	8.91	6.11	6.52	**0.0053**	22.13	22.49	21.62	**0.0000**
Resting in a lying position [h/week]	4.77 ± 5.64	5.66 ± 6.25	5.60 ± 7.79	0.2603	6.28 ± 7.93	4.22 ± 5.51	3.96 ± 4.98	**0.0299**
Sleep time [h]	6.61 ± 1.02	6.62 ± 1.08	6.70 ± 1.14	0.8774	6.65 ± 1.09	6.83 ± 1.21	6.75 ± 1.28	0.3431
Sleep disturbances [score 11–44]	16.87 ± 4.17	15.81 ± 4.56	17.38 ± 4.22	**0.0011**	18.74 ± 4.99	18.02 ± 4.72	18.50 ± 4.84	0.4308
Passive rest at home [h/week]	19.66 ± 10.24	15.48 ± 15.49	6.39 ± 12.75	**0.0016**	15.66 ± 8.89	14.59 ± 7.21	16.44 ± 8.25	0.1152
Low-intensity recreational activity [% of persons]	66.33	52.74	62.16	**0.0001**	62.50	63.39	71.00	0.4676
Moderate-intensity recreational activity [% of persons]	32.32	39.94	40.00	**0.0003**	20.83	30.82	30.30	0.7640
Vigorous-intensity recreational activity [% of persons]	13.25	18.43	17.07	0.4879	10.43	10.07	10.89	0.1852
Daily smokers [% of persons]	49.74	40.00	36.96	**0.0301**	43.55	35.33	35.14	0.2379
Caffeine consumption [mg/day]	297.3 ± 134.3	303.6 ± 157.4	307.6 ± 121.2	0.8134	358.6 ± 149.1	307.3 ± 132.9	366.0 ± 187.7	**0.0002**
Getting drunk once a month or more often [% of persons]	27.33	10.12	26.09	**0.0000**	6.82	0.43	0.00	**0.0003**
Preference for salty foods [% of persons]	39.39	44.35	45.65	0.4038	50.82	47.84	54.95	0.1050
Preference for fatty foods [% of persons]	38.11	29.13	45.65	**0.0259**	37.70	36.21	41.44	0.1505
Diet quality assessment [score 21–84]	58.49 ± 4.37	60.95 ± 5.08	59.35 ± 4.72	**0.0000**	60.52 ± 5.05	61.39 ± 4.89	61.44 ± 5.55	0.2539

Bold means statistically significant value, $p < 0.05$.

The highest ratings of the effect of fatigue, no options for rest and negative life-style indicators, both in the male and in the female studied population, were noted in physical workers, and the highest corresponding values were noted in men exposed to mixed workload and women exposed to mental workload.

In the male group, WAI values differed depending on work content; the highest rating was noted in workers exposed to mixed workload, and the lowest corresponding values were noted in mental workers. Table 4.2 presents WAI values and individual WAI elements values in female and male groups. Similar values were obtained in the female group; however, they were statistically insignificant. The work content statistically significantly differentiated "the number of current diseases diagnosed by a physician". The lowest rating was found in mental workers and those who reported being on "sick leave during the past year", as well as those with "estimated work impairment due to diseases", "own prognosis of WA two years from now" and "mental resources". These WAI components got the lowest rating in female physical workers, although the differences, as compared with those noted for other work contents, were statistically insignificant. The lowest WAI rating was obtained in the group of female mental workers with a different work content, "WA with respect to the mental demands". The work content did not differentiate "current WA compared with the lifetime best" and "WA with respect to the physical demands" or "WA in relation to job demands".

Multiple regression stepwise analysis was applied to indicate the sets of variables being predictors of WAI components and overall WAI. Table 4.3 presents the results obtained in the male group, whereas Table 4.4 presents the corresponding values obtained in the female group. The tables present standardized regression coefficients which, in the case of stepwise regression analysis, allow to compare the significance of predictors in the model (since they eliminate the effect of unequal scales) as well as multivariate regression coefficients (R) and determination coefficients (R^2).

The models obtained from stepwise regression analysis explained the highest percentage of WAI variation in mental workers (45.4%). A lower percentage of this value was noted in workers exposed to mixed load and in physical workers (43.2% and 39.6%, respectively). In the female group, the corresponding percentages of explained variations found for "WA with respect to the physical demands", "estimated work impairment due to diseases" and "WA in relation to the demands of the job" were lower and, depending on the work content, amounted to 39.6%, 35.9% and 25.3%, respectively. In the male group, the highest percentages of the explained variation were found for "WA with respect to the physical demands", "estimated work impairment due to diseases" and "WA in relation to the demands of the job". Conversely, the models obtained in the female group explain to the highest extent such parameters as "own prognosis of work ability two years from now", "current work ability compared with the lifetime best" and "estimated work impairment due to diseases". Both in the male and in the female group, the lowest percentages of variation explained "the number of current diseases diagnosed by a physician" and "sick leave during the past year".

Age and/or seniority were negatively correlated with WA. Only in the women who usually worked in exposure to mental or mixed load, such parameters as "WA with respect to the mental demands" and "WA in relation to the demands of the job" were correlated neither with age nor with seniority.

TABLE 4.2
WAI Values and Individual WAI Components in Female and Male Groups

	Men				Women			
	Work Content				Work Content			
	Mostly Physical N = 393	Mixed N = 230	Mostly Mental N = 46	p	Mostly Physical N = 124	Mixed N = 301	Mostly Mental N = 111	p
Work Ability Index (7–49)	39.31 ± 5.13	41.26 ± 4.73	38.76 ± 5.86	0.0000	39.63 ± 5.00	40.05 ± 5.21	39.59 ± 5.24	0.6288
Categorization of total WAI score [% persons]								
Poor (7–27)	1.53	1.30	2.17	0.0003	0.81	1.99	1.80	0.6229
Moderate (28–36)	24.94	13.91	28.26		27.42	21.26	22.52	
Good (37–43)	52.93	47.83	45.65		50.00	48.17	52.25	
Excellent (44–49)	20.61	36.96	23.91		21.77	28.57	23.42	
Items of WAI								
Current WA compared with the lifetime best (0–10)	7.98 ± 1.54	7.89 ± 1.43	8.15 ± 1.67	0.5044	7.62 ± 1.54	7.68 ± 1.64	7.81 ± 1.60	0.6579
WA with respect to the physical demands (1–5)	4.03 ± 0.78	4.12 ± 0.78	4.02 ± 0.88	0.3401	3.96 ± 0.80	3.91 ± 0.98	3.98 ± 1.09	0.7723
WA with respect to the mental demands (1–5)	4.29 ± 0.77	4.30 ± 0.71	4.38 ± 0.70	0.7500	4.41 ± 0.81	4.34 ± 0.72	4.14 ± 0.81	0.0185
WA in relation to the demands of the job (2–10)	8.19 ± 1.44	8.41 ± 1.37	8.54 ± 1.46	0.0889	8.17 ± 1.44	8.31 ± 1.45	8.19 ± 1.72	0.6133

(Continued)

TABLE 4.2 (Continued)
WAI Values and Individual WAI Components in Female and Male Groups

	Men				Women			
	Work Content				Work Content			
	Mostly Physical N = 393	Mixed N = 230	Mostly Mental N = 46	p	Mostly Physical N = 124	Mixed N = 301	Mostly Mental N = 111	p
Number of current diseases diagnosed by a physician (1–7)	5.36 ± 1.78	5.97 ± 1.24	5.13 ± 1.86	0.0000	5.30 ± 1.98	5.03 ± 1.88	4.53 ± 1.94	0.0075
Sick leave during the past year (1–5)	4.14 ± 1.78	4.42 ± 1.01	3.93 ± 1.31	0.0024	4.35 ± 1.01	4.41 ± 1.06	4.37 ± 1.06	0.8354
Estimated work impairment due to diseases (1–6)	4.93 ± 1.05	5.51 ± 0.84	4.89 ± 1.02	0.0000	5.16 1.05	5.25 ± 0.98	5.32 ± 0.86	0.4732
Own prognosis of WA 2 years from now (1, 4 or 7)	5.80 ± 1.07	5.94 ± 1.26	5.33 ± 1.32	0.0036	6.17 ± 1.43	6.34 ± 1.10	6.43 ± 1.01	0.2055
Mental resources (1–4)	2.90 ± 0.67	3.10 ± 0.65	2.78 ± 0.72	0.0016	2.85 ± 0.67	3.01 ± 0.69	2.93 ± 0.71	0.0875

TABLE 4.3

Results of Backward Regression Analysis: Effects of Individual and Work-Related Factors on WAI and Its Components among Men Performing Mostly Physical or Mixed or Mostly Mental Work

| | Current WA Compared with the Lifetime Best | | | | WA with Respect to the Physical Demands | | | | WA with Respect to the Mental Demands | | | | WA in Relation to the Demands of the Job | | | |
| | Work Content | | | All Men | Work Content | | | All Men | Work Content | | | All Men | Work Content | | | All Men |
	Mostly Physical	Mixed	Mostly Mental		Mostly Physical	Mixed	Mostly Mental		Mostly Physical	Mixed	Mostly Mental		Mostly Physical	Mixed	Mostly Mental	
Multiple correlation coefficients (R)	**0.483**	**0.529**	**0.606**	**0.471**	**0.477**	**0.664**	**0.773**	**0.515**	**0.440**	**0.599**	**0.632**	**0.447**	**0.495**	**0.668**	**0.713**	**0.517**
R-squared (R²)	0.234	0.280	0.367	0.209	0.228	0.440	0.598	0.265	0.193	0.358	0.400	0.200	0.245	0.446	0.509	0.267
Factors								Standardized beta coefficients								
Age	−0.237			−0.241			−0.233		−0.132	−0.273		−0.199				−0.262
Number of years of education									0.214	0.139		0.211				0.115
General seniority		−0.205			−0.350	−0.290		−0.370						−0.304		
Seniority in the workplace																
Seniority in the work post																
Night shift work				−0.103						−0.174		−0.086				
Work time																
Working overtime	0.123			0.113			0.286			0.196	0.336			0.141	0.307	0.074
Energy expenditure	−0.137			−0.095						0.183						
Carrying loads >20 kg																
Periodically heavy physical exertion							−0.336		−0.129				−0.098			

(Continued)

TABLE 4.3 (Continued)
Results of Backward Regression Analysis: Effects of Individual and Work-Related Factors on WAI and Its Components among Men Performing Mostly Physical or Mixed or Mostly Mental Work

| | Current WA Compared with the Lifetime Best | | | | WA with Respect to the Physical Demands | | | | WA with Respect to the Mental Demands | | | | WA in Relation to the Demands of the Job | | | |
| | Work Content | | | | Work Content | | | | Work Content | | | | Work Content | | | |
	Mostly Physical	Mixed	Mostly Mental	All Men	Mostly Physical	Mixed	Mostly Mental	All Men	Mostly Physical	Mixed	Mostly Mental	All Men	Mostly Physical	Mixed	Mostly Mental	All Men
Uncomfortable work post		-0.149				-0.111			-0.160				-0.145	-0.144		
Squatting during work								0.087					0.124			
Leaning during work	0.098				0.100											
Malfunctioning equipment					-0.100		-0.402					-0.081	-0.131			
Continuous walking					-0.116				-0.125				-0.134			
Prolonged sitting	-0.142			-0.104		0.151			-0.141				-0.103			
Awkward posture >50% work time						-0.108								-0.118		
Generally large amount of work																
Imposed pace of work		-0.139												-0.117		
Uneven pace of work																
High repeatability of movements														0.151		
Mineral dust																
Chemical factors				0.078												
Hot																

(Continued)

TABLE 4.3 (Continued)

Results of Backward Regression Analysis: Effects of Individual and Work-Related Factors on WAI and Its Components among Men Performing Mostly Physical or Mixed or Mostly Mental Work

	Current WA Compared with the Lifetime Best				WA with Respect to the Physical Demands				WA with Respect to the Mental Demands				WA in Relation to the Demands of the Job			
	Work Content				Work Content				Work Content				Work Content			
	Mostly Physical	Mixed	Mostly Mental	All Men	Mostly Physical	Mixed	Mostly Mental	All Men	Mostly Physical	Mixed	Mostly Mental	All Men	Mostly Physical	Mixed	Mostly Mental	All Men
Cold																
Inadequate lighting			-0.276													
Humidity						-0.134										
Whole body vibration				0.091												
Noise					-0.123			-0.070								-0.088
Occupational stress									-0.166			-0.121	-0.170			-0.099
Assessment of the physical effort at work				-0.122	-0.225	-0.291		-0.251	-0.137	-0.175	-0.580	-0.098	-0.192	-0.186		-0.183
in relation to own possibilities																
Fatigue after work	-0.147															
Non-professional loads																
Prolonged fatigue		-0.402	-0.474	-0.171		-0.333		-0.209		-0.141		-0.099		-0.279	-0.508	-0.160
Giving up non-professional duties	-0.185						-0.288								-0.321	
Giving up hobbies				-0.113												

(Continued)

TABLE 4.3 (Continued)
Results of Backward Regression Analysis: Effects of Individual and Work-Related Factors on WAI and Its Components among Men Performing Mostly Physical or Mixed or Mostly Mental Work

| | Current WA Compared with the Lifetime Best | | | | WA with Respect to the Physical Demands | | | | WA with Respect to the Mental Demands | | | | WA in Relation to the Demands of the Job | | | |
| | Work Content | | | | Work Content | | | | Work Content | | | | Work Content | | | |
	Mostly Physical	Mixed	Mostly Mental	All Men	Mostly Physical	Mixed	Mostly Mental	All Men	Mostly Physical	Mixed	Mostly Mental	All Men	Mostly Physical	Mixed	Mostly Mental	All Men
No possibility of resting at home	-0.123															
Resting in a lying position					-0.103									-0.108		
Sleep time										0.124				0.137		
Sleep disturbances	-0.159			-0.120												
Passive rest at home										0.173		0.074		0.159		
Low-intensity recreational activity					-0.091	-0.179		-0.119				-0.074		-0.161		-0.112
Vigorous-intensity recreational activity												0.101				0.086
Smoking						0.121										
Caffeine consumption																
Getting drunk																
Preference for salty foods																
Diet quality assessment					0.107				0.099				0.122			

(Continued)

TABLE 4.3 (Continued)

Results of Backward Regression Analysis: Effects of Individual and Work-Related Factors on WAI and Its Components among Men Performing Mostly Physical or Mixed or Mostly Mental Work

	Number of Current Diseases Diagnosed by a Physician				Sick Leave during the Past Year				Estimated Work Impairment Due to Diseases				Own Prognosis of WA 2 Years from Now			
	Work Content				Work Content				Work Content				Work Content			
	Mostly Physical	Mixed	Mostly Mental	All Men	Mostly Physical	Mixed	Mostly Mental	All Men	Mostly Physical	Mixed	Mostly Mental	All Men	Mostly Physical	Mixed	Mostly Mental	All Men
Multiple correlation coefficients (R)	**0.346**	**0.449**	**0.341**	**0.339**	**0.241**	**0.407**	**0.651**	**0.265**	**0.554**	**0.591**	**0.739**	**0.541**	**0.530**	**0.579**	**0.760**	**0.499**
R-squared (R²)	0.120	0.202	0.116	0.115	0.058	0.165	0.424	0.070	0.307	0.349	0.547	0.292	0.281	0.336	0.578	0.249
Factors								Standardized Beta Coefficients								
Age	-0.121			-0.102					-0.309		-0.278	-0.157		-0.313		-0.242
Number of years of education					0.111			0.089								
General seniority												-0.122	-0.191			
Seniority in the workplace	-0.166	-0.141		-0.183									-0.187			-0.132
Seniority in the work post			-0.341				-0.379			-0.154					-0.310	
Night shift work						-0.207			0.239			0.130	0.099	-0.212		
Work time						0.151										
Working overtime										-0.196		-0.071	0.186			
Energy expenditure							0.374							0.166		
Carrying loads >20kg	-0.098			-0.075												0.099
Periodically heavy physical exertion							-0.285									

(Continued)

TABLE 4.3 (Continued)
Results of Backward Regression Analysis: Effects of Individual and Work-Related Factors on WAI and Its Components among Men Performing Mostly Physical or Mixed or Mostly Mental Work

	Number of Current Diseases Diagnosed by a Physician				Sick Leave during the Past Year				Estimated Work Impairment Due to Diseases				Own Prognosis of WA 2 Years from Now			
	Work Content				Work Content				Work Content				Work Content			
	Mostly Physical	Mixed	Mostly Mental	All Men	Mostly Physical	Mixed	Mostly Mental	All Men	Mostly Physical	Mixed	Mostly Mental	All Men	Mostly Physical	Mixed	Mostly Mental	All Men
Uncomfortable work post				-0.082					0.118			0.071				
Squatting during work						0.140			-0.121							
Leaning during work												-0.074				
Malfunctioning equipment										0.164						
Continuous walking																
Prolonged sitting	-0.161					-0.211		-0.123					-0.092			
Awkward posture	-0.127							-0.093								
>50% work time																
Generally large amount of work												0.104				
Imposed pace of work																
Uneven pace of work														0.134		
High repeatability of movements									0.126							
Mineral dust	-0.112															0.086
Chemical factors	-0.122			-0.108					-0.103		-0.385	-0.082				-0.116
Hot																

(Continued)

TABLE 4.3 (Continued)

Results of Backward Regression Analysis: Effects of Individual and Work-Related Factors on WAI and Its Components among Men Performing Mostly Physical or Mixed or Mostly Mental Work

	Number of Current Diseases Diagnosed by a Physician				Sick Leave during the Past Year				Estimated Work Impairment Due to Diseases				Own Prognosis of WA 2 Years from Now			
	Work Content			All Men	Work Content			All Men	Work Content			All Men	Work Content			All Men
	Mostly Physical	Mixed	Mostly Mental		Mostly Physical	Mixed	Mostly Mental		Mostly Physical	Mixed	Mostly Mental		Mostly Physical	Mixed	Mostly Mental	
Cold																
Inadequate lighting															−0.278	
Humidity																
Whole body vibration																
Noise													−0.123			
Occupational stress									−0.165	−0.145		−0.136		−0.194	−0.383	−0.102
Assessment of the physical effort at work in relation to own possibilities													−0.156			−0.089
Fatigue after work	0.226							0.146	−0.116							
Non-professional loads	0.197									−0.152			−0.105		−0.327	
Prolonged fatigue							−0.337			−0.296		−0.122		−0.211		−0.162
Giving up non-professional duties																
Giving up hobbies	−0.170					−0.244		−0.120			−0.287	−0.081			−0.297	
No possibility of resting at home											−0.356				−0.242	

(Continued)

TABLE 4.3 (Continued)
Results of Backward Regression Analysis: Effects of Individual and Work-Related Factors on WAI and Its Components among Men Performing Mostly Physical or Mixed or Mostly Mental Work

| | Number of Current Diseases Diagnosed by a Physician | | | | Sick Leave during the Past Year | | | | Estimated Work Impairment Due to Diseases | | | | Own Prognosis of WA 2 Years from Now | | | |
| | Work Content | | | All Men | Work Content | | | All Men | Work Content | | | All Men | Work Content | | | All Men |
	Mostly Physical	Mixed	Mostly Mental		Mostly Physical	Mixed	Mostly Mental		Mostly Physical	Mixed	Mostly Mental		Mostly Physical	Mixed	Mostly Mental	
Resting in a lying position						0.184		0.093								
Sleep time		0.166														
Sleep disturbances									−0.124		0.331	−0.106				
Passive rest at home													0.093			
Low-intensity recreational activity	−0.103								−0.102			−0.068				
Vigorous-intensity recreational activity	−0.179					−0.168					0.254					
Smoking											−0.281					
Caffeine consumption										0.195	0.381	0.154	0.145			0.090
Getting drunk				−0.092						−0.175		−0.113				−0.084
Preference for salty foods		−0.151							−0.143							
Diet quality assessment								0.077				0.110				

(Continued)

TABLE 4.3 (Continued)

Results of Backward Regression Analysis: Effects of Individual and Work-Related Factors on WAI and Its Components among Men Performing Mostly Physical or Mixed or Mostly Mental Work

	Mental Resources				Work Ability Index			
	Work Content				Work Content			
	Mostly Physical	Mixed	Mostly Mental	All Men	Mostly Physical	Mixed	Mostly Mental	All Men
Multiple correlation coefficients (R)	**0.577**	**0.447**	**0.751**	**0.526**	**0.621**	**0.657**	**0.674**	**0.607**
R-squared (R²)	0.333	0.200	0.530	0.276	0.386	0.432	0.454	0.369
Factors				Standardized Beta Coefficients				
Age					−0.323		−0.288	−0.269
Number of years of education		0.146						
General seniority					−0.127	−0.253		
Seniority in the workplace	−0.087							−0.138
Seniority in the work post				−0.072				
Night shift work	0.170		0.319	0.106	0.112	−0.167		
Work time								
Working overtime								
Energy expenditure					−0.093	0.117		
Carrying loads >20kg								
Periodically heavy physical exertion								
Uncomfortable work post						−0.163		
Squatting during work								
Leaning during work								
Malfunctioning equipment								
Continuous walking								

(Continued)

TABLE 4.3 (Continued)

Results of Backward Regression Analysis: Effects of Individual and Work-Related Factors on WAI and Its Components among Men Performing Mostly Physical or Mixed or Mostly Mental Work

	Mental Resources				Work Ability Index			
	Work Content				Work Content			
	Mostly Physical	Mixed	Mostly Mental	All Men	Mostly Physical	Mixed	Mostly Mental	All Men
Prolonged sitting					−0.111	−0.143		−0.086
Awkward posture >50% work time								
Generally large amount of work								
Imposed pace of work								
Uneven pace of work								
High repeatability of movements								
Mineral dust								
Chemical factors		−0.195		−0.102				
Hot					−0.132			
Cold					0.144			
Inadequate lighting								
Humidity								
Whole body vibration								
Noise								−0.086
Occupational stress			−0.367	−0.224	−0.123		−0.308	−0.095
Assessment of the physical effort at work in relation to own possibilities					−0.206	−0.131		−0.165
Fatigue after work	−0.165							
Non-professional loads								

(Continued)

TABLE 4.3 (Continued)

Results of Backward Regression Analysis: Effects of Individual and Work-Related Factors on WAI and Its Components among Men Performing Mostly Physical or Mixed or Mostly Mental Work

	Mental Resources				Work Ability Index			
	Work Content				Work Content			
	Mostly Physical	Mixed	Mostly Mental	All Men	Mostly Physical	Mixed	Mostly Mental	All Men
Prolonged fatigue		-0.289	**-0.367**	-0.113		-0.381	-0.307	-0.178
Giving up non-professional duties								
Giving up hobbies	-0.215	-0.191		-0.184		-0.142		-0.098
No possibility of resting at home					-0.091			
Resting in a lying position								
Sleep time								
Sleep disturbances				-0.095				
Passive rest at home								
Low-intensity recreational activity	-0.092			-0.077	-0.102	-0.146		-0.110
Vigorous-intensity recreational activity			**0.406**		0.115			
Smoking								
Caffeine consumption	0.123							0.101
Getting drunk	-0.151			-0.224				-0.107
Preference for salty foods								
Diet quality assessment	0.146			0.128	0.095			0.101

Bold means statistically significant value, p < 0.05.

TABLE 4.4

Results of Backward Regression Analysis: Effects of Individual and Work-Related Factors on WAI and Its Items among Women Performing Mostly Physical or Mixed or Mostly Mental Work

	Current WA Compared with the Lifetime Best				WA with Respect to the Physical Demands				WA with Respect to the Mental Demands				WA in Relation to the Demands of the Job			
	Work Content				Work Content				Work Content				Work Content			
	Mostly Physical	Mixed	Mostly Mental	All Women	Mostly Physical	Mixed	Mostly Mental	All Women	Mostly Physical	Mixed	Mostly Mental	All Women	Mostly Physical	Mixed	Mostly Mental	All women
Multiple correlation coefficients (R)	0.505	0.506	0.649	0.482	0.461	0.543	0.514	0.488	0.410	0.410	0.293	0.338	0.431	0.514	0.322	0.403
R-squared (R²)	0.255	0.256	0.421	0.232	0.213	0.295	0.265	0.238	0.168	0.167	0.086	0.114	0.185	0.264	0.104	0.162
Factors																
								Standardized Beta Coefficients								
Age	−0.283	−0.147						−0.093								
Number of years of education				−0.128												
General seniority																
Seniority in the workplace							−0.199	−0.119	−0.290				−0.233			
Seniority in the work post						−0.164						−0.093				
Night shift work																
Work time			0.167													
Additional job	−0.172								−0.164							
Energy expenditure										0.113		0.158				
Generally heavy physical exertion																

(Continued)

TABLE 4.4 (Continued)

Results of Backward Regression Analysis: Effects of Individual and Work-Related Factors on WAI and Its Items among Women Performing Mostly Physical or Mixed or Mostly Mental Work

	Current WA Compared with the Lifetime Best				WA with Respect to the Physical Demands				WA with Respect to the Mental Demands				WA in Relation to the Demands of the Job			
	Work Content				Work Content				Work Content				Work Content			
	Mostly Physical	Mixed	Mostly Mental	All Women	Mostly Physical	Mixed	Mostly Mental	All Women	Mostly Physical	Mixed	Mostly Mental	All Women	Mostly Physical	Mixed	Mostly Mental	All women
Generally large amount of work									-0.177							
Uneven work pace			-0.227								-0.214				-0.192	
High repeatability of movements																
Continuous walking				-0.086									-0.207			
Prolonged standing																
Prolonged sitting										-0.114						
Awkward posture >50% work time	0.214	0.095														
Chemical factors							-0.207	-0.113								
Hot	-0.168															
Cold	-0.179					-0.265							-0.185			
Humidity							0.239									
Noise									0.187			0.104				
Inadequate lighting	0.265															
Occupational stress	-0.224			-0.197	-0.177		-0.170	-0.168				-0.126	-0.279	-0.162		-0.148

(Continued)

TABLE 4.4 (Continued)

Results of Backward Regression Analysis: Effects of Individual and Work-Related Factors on WAI and Its Items among Women Performing Mostly Physical or Mixed or Mostly Mental Work

| | Current WA Compared with the Lifetime Best | | | | WA with Respect to the Physical Demands | | | | WA with Respect to the Mental Demands | | | | WA in Relation to the Demands of the Job | | | |
| | Work Content | | | | Work Content | | | | Work Content | | | | Work Content | | | |
	Mostly Physical	Mixed	Mostly Mental	All Women	Mostly Physical	Mixed	Mostly Mental	All Women	Mostly Physical	Mixed	Mostly Mental	All Women	Mostly Physical	Mixed	Mostly Mental	All women
Assessment of the physical effort at work in relation to own possibilities		-0.147		-0.125	-0.247	-0.261	-0.278	-0.246		-0.139		-0.109		-0.221		-0.176
Fatigue after work																
Prolonged fatigue		-0.222	-0.394	-0.204		-0.136		-0.087		-0.143				-0.192		-0.110
Non-professional loads		-0.100				-0.121								-0.136		
Giving up non-professional duties	-0.218			-0.114			-0.227	-0.101							-0.235	
Giving up hobbies										-0.136		-0.110				-0.110
No possibility of resting at home												-0.110				
Sleep time			-0.207													
Sleep disturbances																
Resting in lying position										0.114		0.084		0.108		

(Continued)

TABLE 4.4 (Continued)

Results of Backward Regression Analysis: Effects of Individual and Work-Related Factors on WAI and Its Items among Women Performing Mostly Physical or Mixed or Mostly Mental Work

	Current WA Compared with the Lifetime Best				WA with Respect to the Physical Demands				WA with Respect to the Mental Demands				WA in Relation to the Demands of the Job			
	Work Content				Work Content				Work Content				Work Content			
	Mostly Physical	Mixed	Mostly Mental	All Women	Mostly Physical	Mixed	Mostly Mental	All Women	Mostly Physical	Mixed	Mostly Mental	All Women	Mostly Physical	Mixed	Mostly Mental	All women
Passive rest at home			0.166													
Low-intensity recreational activity			0.176								0.178	0.085				
Moderate-intensity recreational activity			0.251							0.108						
Smoking							0.121									
Caffeine consumption																
Preference for salty foods																
Preference for fatty foods										−0.117						

(Continued)

TABLE 4.4 (Continued)

Results of Backward Regression Analysis: Effects of Individual and Work-Related Factors on WAI and Its Items among Women Performing Mostly Physical or Mixed or Mostly Mental Work

	Number of Current Diseases Diagnosed by a Physician				Sick Leave during the Past Year				Estimated Work Impairment Due to Diseases				Own Prognosis of WA 2 Years from Now			
	Work Content				Work Content				Work Content				Work Content			
	Mostly Physical	Mixed	Mostly Mental	All Women	Mostly Physical	Mixed	Mostly Mental	All Women	Mostly Physical	Mixed	Mostly Mental	All Women	Mostly Physical	Mixed	Mostly Mental	All Women
Multiple correlation coefficients (R)	0.322	0.404	0.374	0.355	0.263	0.165	0.423	0.151	0.241	0.577	0.680	0.497	0.575	0.490	0.666	0.443
R-squared (R^2)	0.103	0.163	0.140	0.126	0.069	0.027	0.179	0.023	0.058	0.332	0.463	0.247	0.331	0.240	0.443	0.197
Factors								Standardized Beta Coefficients								
Age	−0.305			−0.258												
Number of years of education																
General seniority		−0.260									−0.366	−0.145				
Seniority in the workplace										−0.151					−0.152	
Seniority in the work post																−0.113
Night shift work	0.229															
Work time										−0.141	0.153	−0.102			0.211	
Additional job													−0.321			−0.084
Energy expenditure																
Generally heavy physical exertion																

(Continued)

TABLE 4.4 (Continued)

Results of Backward Regression Analysis: Effects of Individual and Work-Related Factors on WAI and Its Items among Women Performing Mostly Physical or Mixed or Mostly Mental Work

| | Number of Current Diseases Diagnosed by a Physician | | | | Sick Leave during the Past Year | | | | Estimated Work Impairment Due to Diseases | | | | Own Prognosis of WA 2 Years from Now | | | |
| | Work Content | | | | Work Content | | | | Work Content | | | | Work Content | | | |
	Mostly Physical	Mixed	Mostly Mental	All Women	Mostly Physical	Mixed	Mostly Mental	All Women	Mostly Physical	Mixed	Mostly Mental	All Women	Mostly Physical	Mixed	Mostly Mental	All Women
Generally large amount of work																0.096
Uneven work pace															-0.093	
High repeatability of movements											-0.153	-0.092		0.108		
Continuous walking				-0.110		0.118										-0.135
Prolonged standing	-0.156										-0.134		0.232			
Prolonged sitting							-0.225									
Awkward posture >50% work time							-0.190			-0.120			-0.269		-0.316	
Chemical factors																
Hot														0.114		
Cold																
Humidity											-0.219					
Noise														0.152		0.120
Inadequate lighting																
Occupational stress										-0.192		-0.108		-0.179		-0.160

(Continued)

TABLE 4.4 (Continued)

Results of Backward Regression Analysis: Effects of Individual and Work-Related Factors on WAI and Its Items among Women Performing Mostly Physical or Mixed or Mostly Mental Work

| | Number of Current Diseases Diagnosed by a Physician | | | | Sick Leave during the Past Year | | | | Estimated Work Impairment Due to Diseases | | | | Own Prognosis of WA 2 Years from Now | | | |
| | Work Content | | | | Work Content | | | | Work Content | | | | Work Content | | | |
	Mostly Physical	Mixed	Mostly Mental	All Women	Mostly Physical	Mixed	Mostly Mental	All Women	Mostly Physical	Mixed	Mostly Mental	All Women	Mostly Physical	Mixed	Mostly Mental	All Women
Assessment of the physical effort at work in relation to own possibilities					−0.263			−0.093		−0.163	−0.197	−0.185		−0.233		−0.118
Fatigue after work										−0.133						
Prolonged fatigue			−0.217													
Non-professional loads												−0.106		−0.160	−0.169	−0.137
Giving up non-professional duties													−0.326			−0.094
Giving up hobbies													0.232	−0.119		
No possibility of resting at home						−0.118				−0.163		−0.135				
Sleep time															−0.152	
Sleep disturbances	−0.098						−0.259	−0.096	−0.241	−0.123	−0.207	−0.141			−0.250	
Resting in lying position												0.104				

(Continued)

TABLE 4.4 (Continued)

Results of Backward Regression Analysis: Effects of Individual and Work-Related Factors on WAI and Its Items among Women Performing Mostly Physical or Mixed or Mostly Mental Work

	Number of Current Diseases Diagnosed by a Physician				Sick Leave during the Past Year				Estimated Work Impairment Due to Diseases				Own Prognosis of WA 2 Years from Now			
	Work Content				Work Content				Work Content				Work Content			
	Mostly Physical	Mixed	Mostly Mental	All Women	Mostly Physical	Mixed	Mostly Mental	All Women	Mostly Physical	Mixed	Mostly Mental	All Women	Mostly Physical	Mixed	Mostly Mental	All Women
Passive rest at home																
Low-intensity recreational activity	−0.188															
Moderate-intensity recreational activity																
Smoking																
Caffeine consumption											−0.181		−0.190			
Preference for salty foods		−0.209	−0.234	−0.118						0.185		0.125				
Preference for fatty foods			−0.112													

(Continued)

TABLE 4.4 (Continued)
Results of Backward Regression Analysis: Effects of Individual and Work-Related Factors on WAI and Its Items among Women Performing Mostly Physical or Mixed or Mostly Mental Work

	Mental Resources				Work Ability Index			
	Work Content				Work Content			
	Mostly Physical	Mixed	Mostly Mental	All Women	Mostly Physical	Mixed	Mostly Mental	All Women
Multiple correlation coefficients (R)	**0.385**	**0.409**	**0.410**	**0.371**	**0.503**	**0.599**	**0.629**	**0.550**
R-squared (R^2)	0.148	0.167	0.168	0.138	0.253	0.359	0.396	0.302
Factors			Standardized Beta Coefficients					
Age						−0.204		−0.185
Number of years of education								
General seniority		0.149		0.122				
Seniority in the workplace							−0.210	
Seniority in the work post								
Night shift work								
Work time								
Additional job					−0.176			
Energy expenditure								
Generally heavy physical exertion					0.155			
Generally large amount of work								
Uneven work pace						−0.103		
High repeatability of movements								
Continuous walking								
Prolonged standing								
Prolonged sitting								

(Continued)

TABLE 4.4 (Continued)

Results of Backward Regression Analysis: Effects of Individual and Work-Related Factors on WAI and Its Items among Women Performing Mostly Physical or Mixed or Mostly Mental Work

	Mental Resources				Work Ability Index			
	Work Content				Work Content			
	Mostly Physical	Mixed	Mostly Mental	All Women	Mostly Physical	Mixed	Mostly Mental	All Women
Awkward posture >50% work time					-0.181			
Chemical factors								
Hot								
Cold								
Humidity		0.111						
Noise			-0.189				-0.164	
Inadequate lighting								
Occupational stress						-0.212		-0.159
Assessment of the physical effort at work in relation to own possibilities					-0.225	-0.214	-0.172	-0.203
Fatigue after work		-0.136						
Prolonged fatigue		-0.123		-0.124		-0.202	-0.315	-0.177
Non-professional loads				-0.124		-0.154		
Giving up non-professional duties	0.300		-0.234	-0.098	-0.212			-0.123
Giving up hobbies		-0.146		-0.115				
No possibility of resting at home								
Sleep time								
Sleep disturbances							-0.266	-0.113
Resting in lying position								

(Continued)

TABLE 4.4 (*Continued*)

Results of Backward Regression Analysis: Effects of Individual and Work-Related Factors on WAI and Its Items among Women Performing Mostly Physical or Mixed or Mostly Mental Work

	Mental Resources				Work Ability Index			
	Work Content				Work Content			
	Mostly Physical	Mixed	Mostly Mental	All Women	Mostly Physical	Mixed	Mostly Mental	All Women
Passive rest at home								
Low-intensity recreational activity								
Moderate-intensity recreational activity								
Smoking								
Caffeine consumption								
Preference for salty foods								−0.090
Preference for fatty foods	−0.210	−0.107	−0.286	−0.145				

Bold means statistically significant value, $p < 0.05$.

Shift work including night work was negatively correlated with "current work ability compared with the lifetime best" and "WA with respect to the mental demands". In men exposed to mixed load, these components were negatively correlated with "sick leave during the past year", "own prognosis of work ability two years from now" and the overall score obtained for WAI. However, in the male group, work in such system was positively correlated with "mental resources" and overall WAI values obtained in men exposed to physical load. In the group of women exposed mainly to physical load, the parameters corresponding to shift work and night work were positively correlated with "the number of current diseases diagnosed by a physician".

Also positively or negatively correlated with indicators WA were other factors characterizing the organization of work. Working time was positively correlated with "sick leave during the past year" in men exposed to mixed load, whereas in women involved in mental work, this value was positively correlated with "current WA compared with the lifetime best", "estimated work impairment due to diseases" and "own prognosis of work ability two years from now". The negative correlation between working time and "estimated work impairment due to diseases" was only found in women exposed to mixed workload.

Working overtime in men increased the ratings of most of the WA parameters (the correlations were positive) and only decreased the rating of "estimated work impairment due to diseases" in men exposed to mixed workload. In the female group, in turn, working overtime had no significant effect on WA. Conversely, working in another workplace decreased the scores obtained for "current WA compared with the lifetime best", "WA with respect to the mental demands", "own prognosis of WA two years from now" and overall WAI in the female physical workers.

The objective rating of workload (energy expenditure) was negatively correlated only with such parameters as overall WAI and "current WA compared with the lifetime best" in the male physical workers. In the male participants exposed to mixed workload, the correlation between this factor and overall WAI, "WA with respect to the mental demands" and "own prognosis of WA two years from now" was positive. Likewise, the energy expenditure in female group was positively correlated with "WA with respect to the mental demands".

Among the factors related to work post and way of doing work, obtained in the studied group of men, the most frequent predictors in WA models included uncomfortable work post and prolonged sitting at work, whereas in the female group, the most frequent predictors included prolonged walking at work, uneven work pace, prolonged sitting and assuming an awkward posture >50% of work time.

It is of note that also these factors were either negatively or positively correlated with work ability indicators. In the female group, the score corresponding to uneven work pace decreased some WAI parameters only in mental workers.

Exposure to potentially harmful chemical or physical factors in working environment affected only some work ability indicators. Exposure to noise decreased the rating of "work ability with respect to the physical demands" and "own prognosis of work ability two years from now" in the male physical workers, and "WA in relation to the job demands" and overall WAI in all male participants. The exposure to chemical factors was positively correlated with "current WA compared with the lifetime best" in all men, but it was negatively correlated with "the number of current diseases diagnosed by a physician" or "mental resources". In the female group, exposure to chemical agents was negatively correlated only with "WA with respect to the physical demands". Exposure to noise, in turn, was positively correlated with "WA with respect to the mental demands" (in physical workers) and "Own prognosis of work ability two years from now" (in workers exposed to mixed load), but it was negatively correlated with "mental resources" and overall WAI (in mental workers).

Occupational stress was the last factor related to work characteristics. Stress was always negatively correlated, but not under all working conditions and not with every work ability parameter. In the male physical workers, stress was a predictor of "WA with respect to the mental demands" and "WA in relation to the demands of the job", whereas in men exposed to mixed workload, it was the predictor of "own prognosis of WA two years from now" and "mental resources". Moreover, stress was a predictor of overall WAI in the male physical and mental workers and in the entire group of men. In the studied women, the "number of current diseases diagnosed by a physician", "sick leave during the past year" and "mental resources" were the only parameters correlated with stress. Stress was negatively correlated with other WAI components and overall WAI; however, no statistically significant correlations between these WAI components were found in female mental workers.

Not only job characteristics, but also age and seniority were the factors significantly affecting WA. Individual characteristics and subjective ratings of workload were in some cases even more important. In the studied men, the overall period of education was positively correlated to "WA with respect to the mental demands", "WA in relation to the demands of the job", "sick leave during the past year" and "mental resources" (only in men exposed to mixed load). In women, the years of education were rarely found among WA predictors and were positively correlated only in physical workers with "current WA compared with the lifetime best".

The rating of the adjustment of one's own capabilities to physical effort at work and prolonged fatigue was correlated with WAI components, indicating that the levels of physical effort, exceeding the worker's potential and prolonged fatigue were inversely proportional to WA scores (the higher the physical effort and prolonged fatigue levels, the lower the scores). Fatigue after work was not as frequent a predictor of WAI components as prolonged fatigue; moreover, it was not a predictor of overall WAI.

The consequences of fatigue, manifested as giving up hobbies or non-professional duties, and sleep disturbances were always negatively correlated with WAI components, likewise no options for rest at home. Positive correlations, in turn, were found between the time of rest in lying position and passive rest. Sleep length, which rarely was a predictor of some WAI components, was correlated with them positively in men and negatively in women.

The effect of recreational exercise on WA depended on exercise intensity. In the studied men, low intensity of recreational exercise decreased overall WAI and some of its components, whereas high intensity of such exercise increased "WA with respect to the mental demands", "WA in relation to the demands of the job" and "estimated work impairment due to diseases" (in men exposed mainly to mental load), but it decreased the "number of current diseases diagnosed by a physician" and "sick leave during the past year" (in men exposed mixed load). Among women participation in low and moderate recreational exercise was positively correlated with "current WA compared with lifetime best" (in female mental workers) and "WA with respect to the mental demands". However, a negative correlation was found between the "number of current diseases diagnosed by a physician" in female mental workers. Furthermore, a negative correlation was noted between low intensity of recreational exercise and WAI in female physical workers.

In the male group, a positive correlation was noted between daily caffeine consumption and "estimated work impairment due to diseases" (the women exposed to mixed or mental workload), and "own prognosis of WA two years from now" and overall WAI. In the female group, in turn, caffeine consumption was correlated with the same WAI components, but the correlation was negative.

In the male group, the frequency of getting drunk was correlated with four WAI components: "number of current diseases diagnosed by a physician", "estimated work impairment due to diseases", "own prognosis of WA two years from now", "mental resources", and with overall WAI. In the female group, "getting drunk once a month and more often" was rare and had no effect on WA.

In the male group, a positive correlation was found between nutritional habits (not eating too salty foods and sticking to generally well-balanced diet). In the female group, in turn, the preference for salty foods decreased the rating of "number of current diseases diagnosed by a physician", but their preference for high-fat foods increased the rating of "estimated work impairment due to diseases" and "WA with respect to the mental demands" (in those exposed to mixed workload).

DISCUSSION

Work ability was assessed in a large sample of economically active women and men with different job titles, exposed to different workload. The results of work content analysis suggest that the studied men were exposed to higher levels of physical load than women (their energy expenditure was definitely higher). The workload tolerance, however, was higher in men than in women. The rating of WA related to physical effort was higher in the studied men as compared to women. Furthermore, the men more often reported that they had no problems with physical effort or that their tolerance to physical effort was good. The above-mentioned between-subject differences in physical effort tolerance at work did not concern mental workers, since this group reported the lowest levels of exposure to this kind of effort. In men's work, additional difficulties/nuisances occurred more often than in women's work. On the other hand, the mental workload at work for women was greater than for men, which may be indicated by the more frequent involvement of women in doing mental work. The abilities to deal with mental effort were identical in both groups; hence, we can conclude that the tolerance to mental workload was lower in women as compared to their male counterparts. All these kinds of workload resulted in significantly higher levels of work-related fatigue in the studied women, as compared to the fatigue in the male group. Based on this finding, we may conclude that caution should be exercised in eliminating discriminative approaches such as reduction of women's workload, defined by law. On the contrary, workload should always be adjusted to the worker's individual capabilities (Makowiec-Dąbrowska et al. 2009).

The values obtained from work ability assessment in both studied groups were almost identical and the mean value of WAI indicated a good WA in the studied sample. This outcome is probably due to the fact that work ability was assessed by occupationally active people who highly rated prognoses for the next 2 years. The comparison of the values obtained from our study participants with the data obtained in other countries has shown lower work ability ratings in Polish population (Ilmarinen 1999; Monteiro et al. 2006). Such a direct comparison however is unauthorized since WA should be compared only in persons doing the same kind of work.

In the studied sample, WAI levels differed depending on work content. The lowest values were obtained in mental workers (especially male ones), contrary to the values obtained from other studies. Aittomäki et al. (2003) found that lowered work ability was more prevalent among the blue-collar employees than among the white-collar employees of both genders. Such differences may justify the entire context of the socioeconomic situation of occupational groups and the healthy worker effect in physical workers. Both the male and female subsamples included lower and rarely intermediate white-collar workers whose socioeconomic situation was no higher than that of blue-collar workers. Furthermore, among the mental workers, compared to workers with a physical or mixed load, there were significantly fewer people without diseases (assessment of the WAI component of "number of current diseases diagnosed by a physician") and the highest non-professional duties. Our results are thus consistent with those obtained by Schreuder et al. (2008) who suggested that health in the working population depended predominantly on socioeconomic status, and Pohjonen (2001) who stated that hard life situation predicted poor work ability. In addition, health problems are not always the reason for ceasing work if it does not require hard labour. Workers with poor health quit their jobs, especially when they are too burdensome (healthy worker effect). We can assume that this phenomenon is even more related to work ability.

Like many other studies, our research has shown that both WAI and its components are correlated with many factors characterizing study participants as well as their jobs and non-professional load. All these factors can be divided into objectively and subjectively assessed. Objectively assessed are age and seniority as well as all job characteristics except occupational stress. The inclusion of work features as subjectively assessed factors is due to the fact that participants only indicated that such factors exist, but never reported how important these factors were to them. Work-related fatigue, in turn, as well as prolonged fatigue and the consequences of fatigue, non-professional load and lifestyles was subjectively assessed by the participants. Given that WAI mainly includes subjective assessment, we may expect that other subjective ratings will be correlated with this index due to the common source of variance, namely the study participant. This hypothesis is confirmed by our results indicating that the characteristics of study participants will be more useful for WAI prognoses made by men and women exposed to mixed or mental workload. However, in physical workers (men and women) having generally more burdensome jobs, the factors typical for work content were more often encountered among WAI predictors.

Significantly more job characteristics turned out to be important predictors of individual WAI component rating. The factors corresponding to physical load and exposure to dust and noise were specific WA predictors for male physical workers. The predictors that are specific for men exposed to mixed load include the factors corresponding to work organization and working time, while improper lighting is the only predictor in men doing mental work. Shift work including night work, working overtime, energy expenditure, periodically high levels of physical effort and work-related stress were found to be WA predictors in men doing each job. WA predictors specific to physically working women were shift work with night work, working at another workplace, large amounts of work and exposure to low temperatures; in women working in exposure to mixed load, it was generally high physical load, whereas in female mental workers, such predictors included uneven pace of work and exposure to chemical agents. Long-standing, assuming an awkward posture at work and exposure to noise, were the predictors of WA in women doing all kinds of job.

All the work-related factors characterizing work were potential burdens. We can thus expect that such factors will be negatively correlated with WAI or its components. Surprisingly, in some cases, the correlations were positive. Working overtime was correlated with most of the WAI components, which means that WA scores were higher in men with longer work time, indicating better work ability in this group. It is worth noting, however, that working overtime was negatively correlated with "estimated work impairment due to diseases" in men exposed to mixed workload. Energy expenditure during work was (objective measure of workload) also positively or negatively correlated with some WAI components in the male group. Among women, overtime working had no effect on WA, but working in another workplace was negatively correlated with WA indicators. Energy expenditure was positively correlated with "WA with respect to the mental demands". Positive correlations were also found between such a burdensome factor as shift work including night shifts and some WAI components. However, a positive correlation between potentially burdensome factors and WAI components is not an evidence for the protective role of these factors. Therefore, it is incorrect to believe that prolonged working time, night work or high workload would improve WA levels, because the direction of the above-mentioned correlations may also indicate that only people with higher subjective assessments of work ability can work when exposed to higher loads.

Subjective assessment of the adjustment of one's own abilities to physical effort at work is a better predictor of WAI components and overall WAI. The lack of a clear impact of the objectively determined severity of work on the perception of work capacity is probably due to the fact that the worker can only assess how he tolerates this work. It is of note that in most of the studies showing the correlation between physical demands at work and WAI determinants of physical workload, the measurement of physical workload was based on self-report (Tuomi et al. 1997, Van den Berg et al. 2009).

Van den Berg et al. (2009) emphasize, however, that "this assessment technique may lead to spurious results, when subjects with a poor WAI overestimate their physical and mental workload in the workplace relative to those with an excellent WAI" (van den Berg et al. 2009).

The significance of individual responses to work load in WA formation is also manifested by very frequent correlations between prolonged fatigue and WAI components, and overall WAI. Like many other studies, our study has always indicated a negative correlation between fatigue and WA (da Silva et al. 2015; Rostamabadi et al. 2017; Kouhnavard et al. 2018; Garosi et al. 2019). Prolonged fatigue is a cumulative response to excessive load at work and outside of work. The analysis of prolonged fatigue risk indicators presented in another publication, carried out in the same group of workers, showed that they were mostly the same factors that in this analysis turned out to be significantly negatively correlated with WAI (Makowiec-Dąbrowska and Koszada-Włodarczyk 2006). The lack of significant correlations prolonged fatigue with WAI components and overall WAI in women exposed to physical workload results from the fact that prolonged fatigue in this group was statistically significantly correlated only with non-professional load, but not with workload.

The role of work-related stress, being WAI predictor, has been assessed in multiple studies (Tuomi et al. 1991; Kloimuller et al. 2000; Sjögren-Rönkä et al. 2002; Goedhard and Goedhard 2005; Bethge et al. 2009; van den Berg et al. 2009; Yong et al. 2013; Habibi et al. 2014; Martinez et al. 2016; Li et al. 2016). A lower work ability (WAI < 37) was found to be an independent risk factor for high job strain due to high demand and low control (OR = 4.66) and by effort-reward imbalance (OR = 2.88) (Bethge et al. 2009), that WAI values were significantly negatively correlated with the to stress assessment (r = −0.156; p = 0.04) (Habibi et al. 2014), that the work-related stressors affected work ability irrespective of other variables (Martinez et al. 2016) and that work resources (measured based on the worker's self-control of his/her situation at work, awards and social support) were positively correlated to work ability (β = 0.70, p < 0.001), whereas job requirements were negatively correlated to work ability (β = −0.09, p = 0.030) (Li et al. 2016). In our study, occupational stress was negatively correlated with WAI components and overall WAI. The values of standardized regression coefficients were always higher in equations describing the correlations between stress and individual WAI components than overall WAI, and higher for each work content than for the whole groups of men and women. The lack of statistically significant correlations between stress and work ability in women involved in mental work may be due to the fact that stress was measured based on the total of 60 items describing potentially stressogenic requirements and work characteristics. Individual factors such as mental workload or lack of awards may have been responsible for the above finding, as similar results were obtained in the same group when the risk of low work ability was estimated (Makowiec-Dąbrowska et al. 2008).

The results of this study indicate that some of individual factors characterizing lifestyle were negatively correlated with WA. In the male group, these factors included preference for salty foods (evident hypertension risk) (Saneei et al. 2014), whereas in the female group, it was the preference for salty foods, but also for fatty foods. The pathogenetic significance of the latter factor is not obvious. Overall, literature thus far supports that high-fat dairy intake is neutral or protective against obesity development. Conversely, recent studies have indicated that dairy fat intake does not increase the risk of cardiovascular diseases and is associated with a lower risk of obesity and type 2 diabetes (Kratz et al. 2013).

The frequency of getting drunk was the next lifestyle element, negatively correlated with WA in the male group, especially in physical workers. The sole fact of getting drunk was negatively correlated with the period of education (number of years), non-professional load, length of sleep and diet quality, and positively correlated with sleep disturbances, smoking and passive rest forms in leisure time.

A negative effect of sleep disturbances on WA was noted both in the male and the female group while the length of sleep was either positively or negatively correlated to WAI components. The negative effect of sleep disturbances on WA is clear as sleep disturbances in the male group were at the same time positively correlated with shift work including night shifts and the frequency of getting drunk, and in both (male and female) groups, this component was positively correlated with fatigue after work, excessive physical effort compared with the worker's potential, non-professional load as well as stress and prolonged fatigue, which were also the factors negatively correlated with WA.

Participation in low-intensity physical exercise in leisure time was also negatively correlated with WAI components and overall WAI in males exposed to physical workload at work. Low-intensity exercise does not improve physical condition; therefore, it cannot improve WA. Only high-intensity recreational-level physical exercise may contribute to WAI improvement.

CONCLUSIONS

The results obtained from the presented analysis show that the factors determining WAI levels do include not only those describing occupational workload and direct responses to work, but also those describing non-professional load, a possibility of recovery after work and lifestyle. Subjectively assessed factors were correlated with WA more often and better; WA correlations with objective load indicators in professional work were weaker. The studied sample included persons having different jobs and exposed to different loads. The analyses conducted in groups exposed to a relatively uniform load (physical, mixed or mental) allowed the examiners to name specific determinants of WA level and the common factors for all groups. The findings indicate the strongest correlations between WA and subjective rating of effort intensity at work, occupational stress, prolonged fatigue and age. Thus, intervention measures should take advantage of these factors to improve or maintain the desired levels of work ability. It seems that such measures should be focused on the improvement of physical condition (which is expected to improve tolerance to work-related effort), stress reduction and promoting the ability to cope with stress, principles of rational recovery and elimination of anti-health behaviours.

REFERENCES

Aittomäki, A., E. Lahelma, and E. Roos. 2003. Work conditions and socioeconomic inequalities in work ability. *Scand J Work Environ Health* 29(2):159–165.

Bethge, M., F. M. Radoschewski, and W. Müller-Fahrnow. 2009. Work stress and work ability: Cross-sectional findings from the German sociomedical panel of employees. *Disabil Rehabil* 31(20):1692–1699. DOI: 10.1080/09638280902751949.

Cade, J., R. Thompson, V. Burley, and D. Warm. 2002. Development, validation and utilisation of food-frequency questionnaires – A review. *Public Health Nutr* 5(4):567–587. DOI: 10.1079/PHN2001318.

da Silva, F. J., V. E. Felli, M. C. Martinez, V. A. Mininel, and A. P. Ratier. 2015. Association between work ability and fatigue in Brazilian nursing workers. *Work* 53(1):225–232. DOI: 10.3233/WOR-152241.

De Zwart, B. C., M. H. Frings-Dresen, and J. C. van Duivenbooden. 2002. Test-retest reliability of the Work Ability Index questionnaire. *Occup Med (Oxf)* 52(4):177–181. DOI: 10.1093/occmed/52.4.177.

Dudek, B., M. Waszkowska, and D. Merecz. 1999. *Ochrona zdrowia pracowników przed skutkami stresu zawodowego.* Łódź: Instytut Medycyny Pracy.

Garosi, E., S. Najafi, A. Mazloumi, M. Danesh, and Z. Abedi. 2019. Relationship between Work Ability Index and Fatigue among Iranian Critical Care Nurses. *Int J Occup Hyg* 10(3). http://ijoh.tums.ac.ir/index.php/ijoh/article/view/395. Accessed September 18, 2019.

Goedhard, R. G., and W. J. A. Goedhard. 2005. Work ability and perceived work stress. *Int Congr Ser* 1280:79–83 DOI: 10.1016/j.ics.2005.02.051.

Habibi, E., H. Dehghan, S. Safari, B. Mahaki, and A. Hassanzadeh. 2014. Effects of work-related stress on Work Ability Index among refinery workers. *J Educ Health Promot* 3:18. DOI: 10.4103/2277-9531.127598.

Ilmarinen, J. 1999. *Ageing workers in European Union – Status and promotion of work ability, employability and employment.* Helsinki: Finnish Institute of Occupational Health, Ministry of Social Affairs and Health, Ministry of Labour.

Kloimuller, I., R. Karazman, H. Geissler, and I. Karazman. 2000. The relation of age, Work Ability Index and stress-inducing factors among bus drivers. *Int J Ind Ergon* 25(5):497–502.

Kouhnavard, B., G. Halvani, M. R. Najimi, and H. Mihanpour. 2018. The relationship between fatigue and Work Ability Index (WAI) of workers in a ceramic industry in Yazd Province, 2014. *Arch Occup Health* 2(1):63–69.

Kratz, M., T. Baars, and S. Guyenet. 2013. The relationship between high-fat dairy consumption and obesity, cardiovascular, and metabolic disease. *Eur J Nutr* 52(1):1–24. DOI: 10.1007/s00394-012-0418-1.

Krause, N., J. Lynch, G. A. Kaplan et al. 1997. Predictors of disability retirement. *Scand J Work Environ Health* 23(6):403–413. DOI: 10.5271/sjweh.262.

Li, H., Z. Liu, R. Liu, L. Li, and A. Lin. 2016. The relationship between work stress and work ability among power supply workers in Guangdong, China: A cross-sectional study. *BMC Public Health* 16:123. DOI: 10.1186/s12889-016-2800-z.

Makowiec-Dąbrowska, T., and W. Koszada-Włodarczyk. 2006. Przydatność kwestionariusza CIS20R do badania zmęczenia przewlekłego. [The CIS20R questionnaire and its suitability for prolonged fatigue studies] *Med Pr* 57(4):335–345.

Makowiec-Dąbrowska, T., W. Koszada-Włodarczyk, A. Bortkiewicz, E. Gadzicka, and J. Siedlecka. 2009. Czy ciężkość pracy dla kobiet może być taka sama jak dla mężczyzn? [Can heaviness of the work for women be the same as for men?] *Med Pr* 60(6):469–482.

Makowiec-Dąbrowska, T., W. Koszada-Włodarczyk, A. Bortkiewicz et al. 2008. Zawodowe i pozazawodowe determinanty zdolności do pracy. [Occupational and non-occupational determinants of work ability] *Med Pr* 59(1):9–24.

Makowiec-Dąbrowska, T., E. Sprusińska, B. Bazylewicz-Walczak, Z. Radwan-Włodarczyk, and W. Koszada-Włodarczyk. 2000. Zdolność do pracy – nowe podejście do sposobu oceny. [Ability to work – New approach to evaluation methods]. *Med Pr* 51(4):319–333.

Martinez, C. M., A. T. da Silva, L. M. R. Dias de Oliveira, and M. F. Fischer. 2016. Longitudinal associations between stressors and work ability in hospital workers. *Chronobiol Int* 33(6):754–758. DOI: 10.3109/07420528.2016.1167713.

Monteiro, M. S., J. Ilmarinen, and H. R. Corraa Filho. 2006. Work ability of workers in different age groups in a public health institution in Brazil. *Int J OccupSaf Ergon* 12(4):417–427. DOI: 10.1080/10803548.2006.11076703.

Pohjonen, T. 2001. Perceived work ability of home care workers in relation to individual and work-related factors in different age groups. *Occup Med (Lond)* 51(3):209–217.

Radkiewicz, P., and M. Widerszal-Bazyl. 2005. Psychometric properties of Work Ability Index in the light of comparative survey study. *Int Congr Ser* 1280:304–309.

Rostamabadi, A., Z. Zamanian, and Z. Sedaghat. 2017. Factors associated with Work Ability Index (WAI) among intensive care units' (ICUs') nurses. *J Occup Health* 59(2):147–155. DOI: 10.1539/joh.16-0060-OA.

Saneei, P., A. Salehi-Abargouei, A. Esmaillzadeh, and L. Azadbakht. 2014. Influence of Dietary Approaches to Stop Hypertension (DASH) diet on blood pressure: a systematic review and meta-analysis on randomized controlled trials. *Nutr Metab Cardiovasc Dis* 24(12):1253–1261. DOI: 10.1016/j.numecd.2014.06.008.

Schreuder, K. J., C. A. Roelen, P. C. Koopmans, and J. W. Groothoff. 2008. Job demands and health complaints in white and blue collar workers. *Work* 31(4):425–432.

Sjögren-Rönkä, T., M. T. Ojanen, E. K. Leskinen, S. Tmustalampi, and E. A. Mälkiä. 2002. Physical and psychosocial prerequisites of functioning in relation to work ability and general subjective well-being among office workers. *Scand J Work Environ Health* 28(3):184–190. DOI: 10.5271/sjweh.663.

Tavakoli-Fard, N., S. A. Mortazavi, J. Kuhpayehzadeh, and M. Nojomi. 2016. Quality of life, work ability and other important indicators of women's occupational health. *Int J Occup Med Environ Health* 29(1):77–84. DOI: 10.13075/ijomeh.1896.00329.

Tuomi, K., L. Eskelinen, J. Toikkanen et al. 1991. Work load and individual factors affecting work ability among aging municipal employees. *Scand J Work Environ Health* 7(Suppl 1):128–134.

Tuomi, K., J. Ilmarinen, A. Jahkola, L. Katajarinne, and A. Tulkki. 1994. Work Ability Index. In: *Occupational Health Care No. 19*, ed. S. Rautaoja. Helsinki: Institute of Occupational Health.

Tuomi, K., J. Ilmarinen, R. Martikainen, L. Aalto, and M. Klockars. 1997. Aging, work, life-style and work ability among Finnish municipal workers in 1981–1992. *Scand J Work Environ Health* 23(Suppl 1):58–65.

Van den Berg, T. I., L. A. Elders, B. C. de Zwart, and A. Burdorf. 2009. The effects of work-related and individual factors on the Work Ability Index: A systematic review. *Occup Environ Med* 66(4):211–220. DOI: 10.1136/oem.2008.039883.

Vercoulen, J. H. M. M., C. M. A. Swanink, J. F. M. Fennis, J. M. D. Galama, J. W. M. van der Meer, and G. Bleijenberg. 1994. Dimensional assessment of chronic fatigue syndrome. *J Psychosom Res* 38(5):383–392. DOI: 10.1016/0022-3999(94)90099-x.

Yong, M., M. Nasterlack, R. P. Pluto, S. Lang, and C. Oberlinner. 2013. Occupational stress perception and its potential impact on work ability. *Work* 46(3):347–354. DOI: 10.3233/WOR-121556.

Part II

Work Ability and Age

5 Age-Related Physiological Changes Influencing Work Ability

Tomasz Kostka and Joanna Kostka
Medical University of Lodz

CONTENTS

AGE-RELATED CHANGES IN THE CARDIOVASCULAR SYSTEM

Ageing of the cardiovascular system results in decreased general performance or aerobic fitness (aerobic capacity), which is a measure of maximum oxygen absorption (VO_2max). The minimal level of aerobic capacity, necessary for independent life, ranges from 13 to 14 ml O_2/kg/min. Around the 25th year of age, the VO_2max starts to decline by about 10% per decade (Blair et al. 1995). The decline of aerobic capacity is due to the decrease in maximum values corresponding to heart rate (HR) (approx. 6–10 beats/minute/decade) and maximum stroke volume, and a smaller increase in the maximal arteriovenous oxygen difference (AVd). The overall cardiovascular risk increases with age and depends on genetic factors, age, sex, concomitant diseases, environmental factors, lifestyle and occupational status (Sołtysik et al. 2019).

The training aimed at aerobic capacity improvement is an essential element of cardiac prevention and rehabilitation in older people. The effect of physical training includes changes mitigating the impact of ageing on the cardiovascular system, such as reduction of HR at rest and during exercise, an increased elasticity of ventricles and prolongation of the early phase of ventricular filling, an increased strength of stroke volume and contraction of the cardiac muscle, a reduced stiffness of arteries and lower afterload, an improvement of the blood supply and an increase in the threshold for myocardial ischemia (Spirduso et al. 2005). It is estimated that systematic physical activity decreases the coronary risk by approximately 50%–60%. In addition, physical training improves physical performance and delays the decline in VO$_2$max by about 10–20 years for physically active persons, compared with those living a sedentary lifestyle. This is especially important, because even a slight increase in VO$_2$max (3–4 ml/kg/min) may extend the period of physical fitness and independence by 6–7 years. The effect of physical training on aerobic capacity is similar (in relative terms) in older and younger individuals (Nelson et al. 2007).

AGE-RELATED CHANGES IN THE RESPIRATORY SYSTEM AND IMMUNITY

Ageing contributes to the impairment of numerous physiological functions and immune responsiveness of the respiratory system (Sharma and Goodwin 2006). Although the total lung capacity remains relatively unchanged, ageing is associated with a decreased vital capacity and an increased residual volume which impairs physiological breathing capacities (Timiras 2003). The age-related decline in respiratory system function accompanied by ageing of the skin, atrophy of the urinary tract and decreased gastric acidity attenuates local defences against infections. The numbers of circulating immune cells and immunoglobulin levels are relatively unchanged. Nevertheless, changes in cell activity (especially T-lymphocyte function including, among others, increases in CD4$^+$/CD8$^+$ ratio and the proportion of memory cells with a concomitant decrease in naive T lymphocyte count) cause a decline, in both cell-mediated immunity and antibody response to immunogen. All of the aforementioned impairments result in a decreased resistance to infections. Therefore, ageing people are at increased risk of developing bronchitis, pneumonia, influenza, herpes zoster and infections of the urinary tract.

Prevention of infections in ageing population includes isolation (limiting visits during periods of increased incidence), hygienic behaviours, a healthy lifestyle, an appropriate nutrition, regular physical exercises and vaccinations against influenza, pneumococcal infection or tetanus. Other measures that should be applied, especially in hospitalized and institutionalized older persons, include respiratory prophylactics, prevention of aspiration and prevention of urinary tract infections by limiting, among others, indications for catheterization.

One of the consequences of many respiratory diseases is prolonged immobilization, which, especially in older people, has a negative impact on the majority of systems and functions of the body, necessary to maintain functional independence while performing activities of daily living. Infections in older nursing home residents increase the probability of recurrent falls (Pigłowska et al. 2013). In extreme cases, frequent falls may contribute to the patient's death. The impairment of the circulatory, respiratory and musculoskeletal system function resulting from this condition leads to a decrease in performance (Chan and Welsh 1998). Immobilization impairs tolerance to glucose and orthostatic tolerance. It can also cause life-threatening complications of the respiratory system (increased secretion from the bronchial tree or retention of mucous causing a decline in immunity and, in consequence, development of pneumonia). Prolonged immobilization increases the risk of thrombophlebitis and bed sores.

Over several last decades, patients were immobilized due to multiple diseases. It is acknowledged that the adverse consequences of hypokinesia are often more severe than the consequences of the disease itself. Given this fact, exercise can be considered to be the basis of all the activities recommended to aged patient regardless of their health status and physical fitness level.

AGE-RELATED CHANGES IN THE MUSCULOSKELETAL SYSTEM

Ageing in the musculoskeletal system is associated with a number of structural and functional changes. These include a decrease in lean body mass, increased percentage of body fat, reduction of skeletal muscle mass and strength (sarcopenia), decrease in the flexibility of muscles, tendons and ligaments, and decreased bone density, structure and strength. A common problem encountered by older adults is the presence of degenerative changes in the locomotor system (osteoarthritis). The associated pain and limitation of the range of motion significantly reduce involvement in physical exercise in senior population. However, the use of appropriately selected physical exercises is the most successful rehabilitation approach, in both osteoarthritis and sarcopenia. The exercises may alleviate musculo-articular pain and enable proper physical functioning.

The function of skeletal muscles becomes gradually attenuated with advancing age. The age-related deterioration of neuromuscular function is mainly due the loss of muscle mass, changes in muscle quality and neuromuscular control, and increased infiltration of muscles by adipose tissue. The majority of studies investigating age-related changes in muscle function have assessed muscle mass and strength (force). The age-induced loss of muscle strength is usually depicted as linear or curvilinear, accelerating after the fifth decade of life (Metter et al. 1997). The role of muscle strength in maintaining physical capacity in patients with limited mobility has been widely evidenced by research results (Puthoff et al. 2008). In recent years, interest has been increasingly focused on muscle power and muscle shortening velocity (Bottaro et al. 2007). Muscle shortening velocity and hence, the ability to develop power, are closely related to the proportion of type II fibres. It has been reported that although muscle strength and power are strongly correlated with each other, they can differently determine functional abilities. Many activities of daily living require the capacity to perform short, relatively intensive exercises which require generation of appropriate muscle power, e.g. both force and some speed of movement. In studies comparing the relationships of strength and muscle power to functional capacity, muscle power almost always appears to be a stronger determinant (Puthoff et al. 2008). The major muscles, necessary to maintain independence during the activities of daily living, are the lower limb muscles. Knee extensor muscles determine the ability to carry out such basic tasks as maintaining a standing position, walking, climbing and descending stairs, as well as rising from a sitting position. Walking speed, which is an indicator of independence in daily living activities, especially in the elderly, is dependent on the strength of the knee extensor muscles (Bohannon 2008).

Bone structure, histology and function become gradually weakened with advancing age, more frequently in women. Osteoporosis is complicated by fractures, and further consequences of this condition include increased institutionalization, mortality and poor quality of life. Densitometry (DXA) remains the "gold standard" in diagnosing osteoporosis. Low-energy hip fractures in women over the age of 50 years and men over the age of 65 years as well as every past low-energy fracture in other major locations (after excluding other causes), with already existing osteopenia (DXA, $T - score < -1.5$), are obvious bases for diagnosing osteoporosis and indications for early initiation of comprehensive treatment (Lorenc et al. 2017).

Weakening of the musculoskeletal system, accompanied by changes in the cardiovascular system and sense organs, coexisting diseases and drug abuse, leads to an increased probability of falls. Due to serious consequences, falling is one of the most critical issues of geriatric medicine. The most common effects of falls are fractures including health- and life-threatening femoral neck fracture.

AGE-RELATED CHANGES IN THE NERVOUS SYSTEM

Both somatic and cognitive disorders determine the functional status of older persons (Agüero-Torres et al. 2002). The ageing brain and spinal cord lose neurons and the supporting neuroglial cells (atrophy). Neurodegenerative changes (β-amyloid storage, hyperphosphorylation of tau protein, abnormal accumulation of lipofuscin, neurofibrillary tangle formation) further impede normal brain function. Vascular age-related pathophysiologic changes (calcification, fibrosis, hyalinosis of blood vessels) and purely pathologic conditions (amyloid angiopathy, microhaemorrhages) contribute to neuronal destruction. Likewise, peripheral nerve damage becomes evident (loss of mitochondria, degeneration of myelin sheaths, reduction of axonal length). The loss of myelin slows the conduction of peripheral nerve impulses and reaction time. The decline in brain function usually starts with a loss of short-term memory and the ability to acquire new information. Loss of memory, thinking, verbal abilities, the ability to process information and the ability to perform tasks are characteristic features of age-related cognitive decline.

Due to their increasing occurrence with age, several debilitating neurological and psychiatric disorders contribute to further deterioration of neuropsychological functioning. Stroke is one of the leading causes of disability in older people (Kostka et al. 2019). About 30% of people who survive a stroke become disabled and require assistance. Other important neurodegenerative disorders include Alzheimer's disease and different types of dementia, Parkinson's disease, depression and delirium.

AGE-RELATED CHANGES IN SENSORY ORGANS

Normal ageing is associated with changes (loss) of sensory abilities, such as vision, hearing, smell, taste and peripheral sensation (touch, temperature).

Visual acuity becomes impaired as we age. A decrease in the ability to change the shape of the lens (accommodation) to focus on near or distant objects and to adapt to light (presbyopia) becomes evident usually after the 45th year of life (Timiras 2003).

Several diseases that alter vision, more frequently experienced in advanced age include the following:

1. Cataract: opacification of (normally clear) lens which can be either unilateral or bilateral.
2. Age-related macular degeneration (AMD): central vision loss due to accumulation of deposits in the macula. There are two types of macular degeneration, namely the dry type and the wet type. It becomes the most common cause of blindness in advanced age.
3. Glaucoma: amplification of intraocular pressure that may cause atrophy of the optic nerve.
4. Diabetic and hypertensive retinopathy: damage of the retina due to a long-term and poorly controlled diabetes or hypertension.

The above changes considerably affect functional and work abilities as well as patient safety (impaired reading, difficulty in walking or driving, problems with navigating objects, etc.).

Age-associated hearing loss (presbycusis) affects men more often than women and is intensified in exposure to higher frequency sounds. The diseases that become more frequent as people age, further altering their hearing, include, among others, tinnitus (ringing in the ears) or Meniere's disease. Sensory changes in hearing can severely impact functional abilities of aged persons as well as their communication skills. Fast speech, speech with reverberation and background noise further aggravate hearing deficit.

The senses of smell and taste deteriorate with ageing. This fact may be important for safety reasons, e.g. when the patient is unable to smell smoke or a gas leak. These changes may become more evident during early stages of Alzheimer's or Parkinson's disease.

Peripheral sensation declines slightly with age, and peripheral neuropathy is one of the most common neurological disorders encountered in older population. Both superficial sensation (touch, temperature, pain) and deep sensation of vibration and proprioception decline with ageing. Impaired feeling of hot or cold temperatures and inability to recognize position are the examples of potential risk associated with attenuated peripheral sensation.

OTHER AGE-RELATED CHANGES AFFECTING FUNCTIONING IN ADVANCED AGE

Ageing is accompanied by profound changes in hormone secretion. Gonadal hormone secretion decreases with age, in both women and men. The decrease in the levels of circulating anabolic hormones, namely dehydroepiandrosterone (DHEA), growth hormone (GH), insulin-like growth factor I (IGF-I) and testosterone, is accompanied by unchanged or elevated cortisol levels. The decline in gonadal hormones is related to an increased secretion of pituitary gonadotropins, namely the follicle-stimulating hormone (FSH) and the luteinizing hormone (LH). The changes in thyroid hormones are equivocal due to various stimuli and disease-related instability.

The morphology and function of kidneys become affected with advancing age in many older individuals who meet the criteria of chronic kidney disease (decreased glomerular filtration rate). Decreased urinary muscle tone, urinary retention, oestrogen deficiency with atrophic changes in women and prostate gland enlargement in men are the factors contributing to urinary incontinence and frequent urinary tract infections.

Decreased motility of the large intestine is the major cause of frequent constipations. Deteriorating function of pelvic muscles and neuromuscular control results in faecal incontinence. Pancreatic and liver functions remain relatively unchanged with age; however, the incidence of many diseases (e.g., bile duct stones) increases considerably in ageing population (Timiras 2003).

Age-related changes in the skin and its appendages include decreased thickness, increased dryness and roughness, as well as increased sensibility and propensity for pressure sores.

TESTS AND SCALES USED IN GERIATRIC ASSESSMENT OF FUNCTIONAL ABILITIES

Comprehensive geriatric assessment (CGA) is a multifaceted, integrated diagnostic process, aimed at determining the scope of deficits in well-being (according to the WHO definition of health), setting priorities and rehabilitation needs, and the opportunities to provide further treatment/rehabilitation/care (home, home care, nursing facility, hospital ward) (Bernardi et al. 2003).

CGA includes an assessment of functional status, physical health (including nutritional status and the risk of developing pressure ulcers), mental function (cognitive function, mood) and socio-environmental functioning. It allows to specify aged persons' ability to function and to address the health-related, psychological and social needs of the elderly population (Devons 2002). Team and interdisciplinary approach should be a standard in the CGA. Due to the large diversity of the problems encountered by patients, a team of specialists representing various fields (physicians, nurses, physical therapists, psychologists, nutritionists and social workers) should be engaged in the process of assessment. Each of team members should make an assessment of the patient with the help of their profession-specific tools (Ekdahl et al. 2015).

From work ability perspective, assessment of the level of functional and rehabilitation potential of an older person is the most important issue. However, interpretation skills and the knowledge about other elements of CGA affect the process of physiotherapy programming and guidelines for proceeding with an older patient. For example, a patient with cognitive deficits or mood disorders in a similar functional state will require a different approach. Likewise, a patient who lives alone requires implementation of procedures which are different from those applied for patients living with their family members who assist them in daily life.

A group of older people is very diverse in terms of health and functional abilities. Therefore, age alone is not an indication for CGA implementation. In Poland, there is a screening tool being used, including whether complete CGA application is necessary, namely the Vulnerable Elders Survey (VES-13) scale (Saliba et al. 2001), which is recommended for use by service providers on the basis of direct interview or over the phone. Scores ≥ 3 indicate a risk of health deterioration and is an indication for implementation of full CGA.

There are no specific guidelines regarding the use of specific tools to implement CGA. It is important, however, to assess the patient in terms of functional status, physical health, mental function and socio-environmental issues.

Assessment of the functional status determines the functioning and independence of senior citizens in the world around us. The methods most commonly used in functional assessment scales are assessment of the basic activities of daily living, or everyday functioning (Activities of Daily Living – ADL) and evaluation of complex (instrumental) activities of daily living (Instrumental Activities of Daily Living – IADL). ADL scale contains six questions about the patient's degree of independence during daily activities, which include personal hygiene, dressing, using the toilet, faecal and urinary continence (having control over urine and stool), feeding and basic movement abilities (standing up from a bed and sitting down, and getting up from a chair). A low score according to the scale (0–2 points) demonstrates inability to function independently and indicates the need of assistance from other people in various activities of daily living, 3–4 points indicate moderate incapacity, while 5–6 points indicate the patient's independence in ADL performance (Katz et al. 1963).

IADL scale assesses eight operational parameters of instrumental abilities of daily living: using a phone, shopping, food preparation, laundry, washing, using means of transport, taking medications and handling finances. Therefore, it assesses functional independence in the surrounding world. The number of points is directly proportional to the respondent's autonomy (Lawton and Brody 1969).

The Barthel scale is a very similar research tool, used to assess the level of functional abilities. It assesses activities of daily living such as eating, moving, maintaining personal hygiene, dressing up and incontinence. The fewer the points the patient receives, the more the assistance be needs. The results should be interpreted in the following way: 86–100 points – the patient is "relatively independent"; 21–85 points – the patient is "partly dependent"; 0–20 points – the condition is "serious" and the patient is "totally dependent" (Mahoney and Barthel 1965).

The EASY-Care questionnaire is another research tool for comprehensive evaluation of the patient's functional status. It enables analysis of the sociomedical needs in residential care settings. The EASY-Care contains questions assessing the patients' vision, hearing and food chewing ability. It also includes assessment of self-esteem-health, well-being and efficiency in terms of performance of the ADL and IADL. EASY-Care also assesses the degree of cognitive impairment and contains questions about the respondent's individual needs (Philip et al. 2014).

Mobility assessment is of major importance in older subject. The Timed "Up and Go" test (TUG) assesses the basic functions of everyday life such as change of position (moving from sitting to standing and vice versa), walking a distance of 3 m, returning and walking back and returning to the sitting position. It is also used to test dynamic balance and to assess the risk of falling. The time is measured using a stopwatch. The time longer than 14 seconds indicates an increased risk of falling (Podsiadlo and Richardson 1991).

The balance and gait scale is a more complex tool (Tinetti test, POMA – Performance-Oriented Mobility Assessment). The final assessment is based on scores obtained from two components: balance and gait. The maximal score for balance is 16 points, whether the maximal score for the gait is 12 points. The overall maximum value is 28 points. The result below 26 points indicates a problem, under 19 points the risk of falls of the patient rises 5 times (Tinetti 1988).

The Functional Reach Test has been developed to test balance. It measures the extreme range of motion in the upper limbs in patients assuming a standing position without support (Duncan et al. 1992).

Test Berg Balance Scale (BBS) is also used to assess static and dynamic balance, the ability to transfer as well as the selection of appropriate assistance to facilitate walking (Berg et al. 1992). It consists of 14 simple activities measured using a five-point scale ranging from of 0 to 4. The maximum score that can be obtained by the patient is 56 points.

The 14 activities measured using BBS are given as follows:

1. Sitting unsupported _____
2. Change of position: sitting to standing _____
3. Change of position: standing to sitting _____
4. Transfers _____
5. Standing unsupported _____
6. Standing with eyes closed _____
7. Standing with feet together _____
8. Tandem standing _____
9. Standing on one leg _____
10. Trunk rotation (feet fixed) _____
11. Retrieving objects from floor _____
12. Turning 360° _____
13. Stool stepping _____
14. Reaching forward while standing

INTERPRETATION

Interpretation is given as follows: 41–56 points indicated that the patient stands alone; 37–40 points indicated navigating with a cane or other support; 21–36 points indicated movements guided by walking canes; and ≤20 points indicated that it is recommended to use a wheelchair.

The 6-minute walk test (6-Minute Walk Test – 6MWT) (Guyatt et al. 1985). The original purpose of the test was to evaluate exercise tolerance in patients with respiratory and circulatory system diseases. Currently, the test is also used in other areas to evaluate exercise tolerance and efficacy of therapies. The results are based on the distance (in meters) walked by the patient within 6 minutes.

TABLE 5.1

Norms (Cut-off Criteria for Frailty) for Handgrip (Fried et al. 2001)

Women	Men
>17 kg for BMI ≤ 23	>29 kg for BMI ≤ 24
>17.3 kg for BMI 23.1–26	>30 kg for BMI 24.1–26
>18 kg for BMI 26.1–29	>30 kg for BMI 26.1–28
>21 kg for BMI > 29	>32 kg for BMI > 28

EVALUATION OF MUSCLE STRENGTH

There is a variety of methods assessing muscle strength. Some of them involve using a large and heavy equipment; therefore, the application of these approaches is limited to research-work in specialist centres.

In clinical practice, the Lovett test has been designed for subjective assessment of muscle strength (Medical Research Council 1943). It distinguishes between the following degrees of muscle strength: 0° – no active muscle contraction during attempted movement; 1° – slight, visible active muscle contraction; 2° – movement performance in full range of motion with assistance; 3° – active contraction in full range of motion; 4° – the ability to perform active movement with submaximal resistance; 5° – active motion against gravity with full resistance.

Dynamometer test is the most commonly used approach to assess strength of muscles during handgrip. Research results indicate that handgrip strength is correlated with the strength of other muscles. The advantage of this measurement is that it is relatively cheap and easy to apply. Therefore, it can be widely used for screening. Standard values for handgrip strength are presented according to sex and BMI (Table 5.1).

Different test batteries are used for functional status assessment in older patients. The Short Physical Performance Battery (SPPB) (Guralnik et al. 1994) is probably the most popular test used by researchers. It assesses functional capacity of this group of patients during performance of daily activity components, such as balance, gait speed and rising from a sitting position. Each part of the test is preceded by explanation and demonstration. Maximum 4 points can be obtained for proper execution of tasks in each category. The maximum score that can be obtained from the whole set of SPPB is 12 points.

The Senior Fitness Test (Fullerton test), developed by Rikli and Jones (1999), is another test measuring the functional status of older patients. It consists of six trials evaluating different functional elements:

1. *Bending the forearm (arm curl)* measures the strength of the upper limbs. The score is based on the number of repetitions of forearm flexion with weight (5 lb-2.27 kg-for women; 8 lb-3.63 kg for men) in the 30 seconds
2. *Getting up from a chair in 30 seconds* (30-Second Chair Stand). The strength of lower limbs is assessed based on the number of times one rises up from a chair in 30 seconds.
3. *The Back Scratch Test* measures flexibility of the upper body.

4. *The Chair Sit-and-Reach Test* measures flexibility of the lower body.
5. *The 8-ft-Up-and-Go Test,* over a distance of 2.45 m, evaluates dynamic balance.
6a. *The 6-Minute Walk Test* – the result is walking distance covered at a time of 6 minutes, or alternatively, if it is not possible to perform:
6b. *The 2-Minute Step Test* – '2 minutes' step in place – the result is the number of upstrokes of the knee at a time of 2 minutes.

The above tests and scales do not cover all the available approaches assessing functional abilities of older patients. Some tests are designed to measure disease-related deficits in groups of specific patients (e.g. after a stroke, Parkinson's disease).

Evaluation of mental function involves assessment of cognitive performance and mood. Based on this evaluation, we can determine the threats and opportunities of cooperation with patients diagnosed with cognitive impairment. Under such circumstances, it is essential to anticipate problems with understanding, concentration and learning new activities, loss of memory or the ability to perform even basic activities of daily living, by implementing adequate measures in accordance with the therapist's or caregiver's advice. In patients with depression, in turn, motivation to perform exercise or undergo therapy can be lower.

Assessment of cognitive function and mental status is based on several tests. Hodkinson's Abbreviated Mental Test Score (AMTS), the Mini-Mental State Examination (MMSE) and the Geriatric Depression Scale (GDS) are the most widely used ones.

AMTS and MMSE assess the degree of functioning in basic mental processes, such as orientation in time and place, remembering, attention and counting, memory recall, language functions and constructive praxis.

Interpretation of the AMTS (maximum 10 points) is given as follows: >6 points – the result within normal values; 4–6 – moderate impairment; 0–3 – heavy impairment (Hodkinson 1972).

Interpretation of the MMSE (maximum 30 points) is given as follows: 27–30 points – the result within normal values; 24–26 – suggestive of mild cognitive impairment without dementia; 19–23 possible light dementia; 11–18 – medium degree dementia; 0–10 points – deep dementia (Folstein et al. 1975).

Fifteen-item GDS consists of 15 questions with Yes or No answers. The results are interpreted in the following way: ≤5 points – no signs of depression; 6–10 points – moderate depression; 11–15 points – severe depression (Yesavage et al. 1983).

In patient evaluation (especially in older patients), importance is increasingly attributed to the quality of life (United Nations 2011). The quality of life can be defined as a multi-dimensional self-esteem status including physical, mental and social well-being. Health-related quality of life (HRQL) is defined as the functional effect of health status on patients.

Euro-Qol 5D is one of our most relevant questionnaires used for evaluation of the quality of life (Kind et al. 1998). It assesses five dimensions of the quality of life. Part of the questionnaire is also a subjective scale of self-health esteem (0–100 points).

REFERENCES

Agüero-Torres, H., V. S. Thomas, B. Winblad, and L. Fratiglioni. 2002. The impact of somatic and cognitive disorders on the functional status of the elderly. *J Clin Epidemiol* 55(10):1007–1012.

Berg, K., S. Wood-Dauphinee, J. I. Williams, and B. Maki. 1992. Measuring balance in the elderly: Validation of an instrument. *Can J Pub Health* (Suppl 2):S7–S11.

Bernardi, D., I. Milan, M. Balzarotti, M. Spina, A. Santoro, and U. Tirelli. 2003. Comprehensive geriatric evaluation in elderly patients with lymphoma: Feasibility of a patient-tailored treatment plan. *J Clin Oncol* 21(4):754.

Blair, S. N., H. W. III Kohl, C. E. Barlow, R. S. Jr Paffenbarger, L. W. Gibbons, and C. A. Macera. 1995. Changes in physical fitness and all-cause mortality: A prospective study of healthy and unhealthy men. *JAMA* 273:1093–1098.

Bohannon, R. W. 2008. Population representative gait speed and its determinants. *J Geriatr Phys Ther* 31(2):49–52.

Bottaro, M., S. N. Machado, W. Nogueira, R. Scales, and J. Veloso. 2007. Effect of high versus low-velocity resistance training on muscular fitness and functional performance in older men. *Eur J Appl Physiol* 99(3):257–264.

Chan, E. D., and C. H. Welsh. 1998. Geriatric respiratory medicine. *Chest* 114(6):1704–1733.

Devons, C. A. 2002. Comprehensive geriatric assessment: Making the most of the aging years. *Curr Opin Clin Nutr Metab Care* 5(1):19–24.

Duncan, P. W., S. Studenski, J. Chandler, and B. Prescott. 1992. Functional reach: Predictive validity in a sample of elderly male veterans. *J Gerontol* 47:M93–M98.

Ekdahl, A. W., F. Sjöstrand, A. Ehrenberg et al. 2015. Frailty and comprehensive geriatric assessment organized as CGA-ward or CGA-consult for older adult patients in the acute care setting: A systematic review and meta-analysis. *Eur Geriatr Med* 6(6):523–540.

Folstein, M. F., S. E. Folstein, and P. R. McHugh. 1975. "Mini-Mental State": A practical method for grading the cognitive state of patients for the clinician. *J Psychiatr Res* 12:189–198.

Fried, L. P., C. M. Tangen, J. Walston et al. 2001. Frailty in older adults: Evidence for a phenotype. *J Gerontol A Biol Sci Med Sci* 56(3):M146–M156.

Guralnik, J. M., E. M. Simonsick, L. Ferrucci et al. 1994. A short physical performance battery assessing lower extremity function: Association with self-reported disability and prediction of mortality and nursing home admission. *J Gerontol A Biol Sci Med Sci* 49(2):M85–M94.

Guyatt, G. H., M. J. Sullivan, P. J. Thompson et al. 1985. The 6-minute walk: A new measure of exercise capacity in patients with chronic heart failure. *CMAJ* 132(8):919–923.

Hodkinson, H. M. 1972. Evaluation of mental test score for assessment of mental impairment in the elderly. *Age Ageing* 1:233–238.

Katz, S., A. B. Ford, R. W. Moskowitz, B. A. Jackson, and M. W. Jaffe. 1963. Studies of illness in the aged: The index of ADL, a standardized measure of biogical and psychosocial function. *JAMA* 185:914–919.

Kind, P., P. Dolan, C. Gudex, and A. Williams. 1998. Variations in population health status: Results from a United Kingdom national questionnaire survey. *BMJ* 316(7133):736–741. https://www.bmj.com/content/316/7133/736 (accessed January 17, 2020).

Kostka, J., M. Niwald, A. Guligowska, T. Kostka, and E. Miller. 2019. Muscle power, contraction velocity and functional performance after stroke. *Brain Behav* 9(4):e01046:1–7. https://onlinelibrary.wiley.com/doi/full/10.1002/brb3.1243 (accessed January 17, 2020).

Lawton, M. P., and E. M. Brody. 1969. Assessment of older people: Self-maintaining and instrumental activities of daily living. *Gerontologist* 9:179–186.

Lorenc, R., P. Głuszko, E. Franek et al. 2017. Guidelines for the diagnosis and management of osteoporosis in Poland: Update 2017. *Endokrynol Pol* 68(5):604–609.

Mahoney, F. I., and D. W. Barthel. 1965. Functional evaluation: The Barthel Index. *Md State Med J* 14:61–65.

Medical Research Council. 1943. *Aids to the investigation of peripheral nerve injuries. War Memorandum No 7.* 2nd ed. London: H.M.S.O.

Metter, E. J., R. Conwit, J. Tobin, and J. L. Fozard. 1997. Age-associated loss of power and strength in the upper extremities in women and men. *J Gerontol A Biol Sci Med Sci* 52(5):B267–B276.

Nelson, M. E., W. J. Rejeski, S. N. Blair et al. 2007. Physical activity and public health in older adults: Recommendation from the American College of Sports Medicine and the American Heart Association. *Circulation* 116(9):1094–1105.

Philip, K. E., V. Alizad, A. Oates et al. 2014. Development of EASY-Care, for brief standardized assessment of the health and care needs of older people: With latest information about cross-national acceptability. *J Am Med Dir Assoc* 15(1):42–46.

Pigłowska, M., J. Kostka, and T. Kostka. 2013. The relationship of respiratory tract infections to falls incidence in nursing home residents. *Pol Arch Intern Med* 123(7/8):371–377.

Podsiadlo, D., and S. Richardson. 1991. The timed "up & go": A test of basic functional mobility for frail elderly persons. *J Am Geriatr Soc* 39:142–148.

Puthoff, M. L., K. F. Janz, and D. Nielson. 2008. The relationship between lower extremity strength and power to everyday walking behaviors in older adults with functional limitations. *J Geriatr Phys Ther* 31(1):24–31.

Rikli, R. E., and C. J. Jones. 1999. Functional fitness normative scores for community-residing older adults, ages 60–94. *J Aging Phys Activity* 7(2):162–181.

Saliba, D., M. Elliott, and L. Z. Rubenstein. 2001. The vulnerable elders survey: A tool for identyfing vulnerable older people in the community. *J Am Geriatr Soc* (49):1691–1699.

Sharma, G., and J. Goodwin. 2006. Effect of aging on respiratory system physiology and immunology. *Clin Interv Aging* 1(3):253–260.

Sołtysik, B. K., J. Kostka, K. Karolczak, C. Watała, and T. Kostka. 2019. What is the most important determinant of cardiometabolic risk in 60–65-year-old subjects: Physical activity-related behaviours, overall energy expenditure or occupational status? A cross-sectional study in three populations with different employment status in Poland. *BMJ Open* 9:e025905. https://bmjopen.bmj.com/content/9/7/e025905 (accessed January 17, 2020).

Spirduso, W., K. Francis, and K. MacRae. 2005. *Physical dimensions of aging.* 2nd ed. Champaign, IL: Human Kinetics.

Timiras, P. S., ed. 2003. *Physiological basis of aging and geriatrics.* 3rd ed. Boca Raton, FL: CRC Press.

Tinetti, M. 1988. Performance-oriented assessment of mobility problems in elderly patients. *J Am Geriatr Soc* 34:119–126.

United Nations Department of Economic and Social Affairs Office of the High Commissioner for Human Rights. 2011. *Current status of the social situation, well-being, participation in development and rights of older persons worldwide.* Department of Economic and Social Affairs Office of the High Commissioner for Human Rights. New York.

Yesavage, J. A., T. Brink, and O. Lom. 1983. Development and validation of a geriatric depression screening scale: A preliminary report. *J Psychiatr Res* 17:37–49.

6 Activities for Supporting Work Ability of Ageing Workers

Joanna Kostka and Tomasz Kostka
Medical University of Lodz

CONTENTS

PHYSICAL ACTIVITY AND NUTRITIONAL RECOMMENDATIONS IN OLDER ADULTS

PHYSICAL ACTIVITY

The levels of physical activity (PA) diminish with age in terms of both volume and intensity. This process is due to changes in lifestyle after the end of professional activity, the growing number of concomitant diseases, lack of motivation, and the insufficient knowledge and awareness about the importance of living an active lifestyle. Meanwhile, the so-called sedentary lifestyle has a negative impact on the majority of systems and functions of the body, and is an important risk factor for non-communicable diseases. The changes resulting from a sedentary lifestyle overlap with the effects of ageing, often aggravating them. Therefore, promoting active lifestyles in older people should be one of the priorities of care in advanced age.

One of the most important factors mitigating the effects of age and promoting "successful ageing" is regular PA (Garber et al. 2011; Nelson et al. 2007). A sedentary lifestyle, so common in older people, has a negative impact on the majority of systems and functions of the body, necessary to maintain independence in performing activities of daily living. There is an inverse relationship between the total energy expenditure (TEE) in middle-aged and older people and overall mortality. Prevention of cardiovascular diseases (CVD) is probably the most important mechanism of action of PA. Regular PA also reduces the risk of developing a number of conditions and diseases associated with age, such as diabetes, some types of cancer (breast cancer, colorectal cancer), osteoporosis, obesity and depression. Regular

PA also helps reduce the risk of falling as well as the incidence of infections, and improves the level of cognitive function and the quality of life.

In younger people, prevention activities are mainly focused on cardiovascular and metabolic diseases. In later years, more emphasis is placed on the prevention of disabilities and ageing-associated diseases (such as sarcopenia or osteoporosis). This finding indicates that health training should include typical strength exercises (ACSM 2009).

The number of physically active older people is gradually increasing. This phenomenon is due to several factors. First, in the developed countries, a steady rise in elderly population is noted. Second, the knowledge on the benefits of regular PA, recommendations from scientific societies and medical service, promotion of PA, improvements in standard of living and favourable conditions for exercise performance contribute to the popularity of an active lifestyle among senior citizens.

The recommendations of the international scientific societies (American College of Sports Medicine, American Heart Association) and the WHO on PA for older people are consistent (WHO 2009; Chodzko-Zajko et al. 2009; Garber et al. 2011; Nelson et al. 2007). Seniors are advised to limit their periods of inactivity and to gradually increase their involvement in PA until they reach the recommended level. PA programs for senior population should include exercises favourably affecting their aerobic capacity, muscle strength and balance, coordination and flexibility.

A program of PA for seniors should contain the following three elements: (1) endurance exercises (walking, running, swimming, cycling, cross-country skiing); (2) 20-minute strength (resistance) exercises; (3) at least 5–10-minute daily workout including stretching, balance and coordination.

Regular PA can bring significant health benefits to people of all ages (Chodzko-Zajko et al. 2009). Older adults should participate in aerobic PA of moderate intensity, at least 150 minutes or 75 minutes per week of vigorous activity (or an equivalent combination of moderate- and vigorous-intensity activity; 1 minute of vigorous intensity exercise equals 2 minutes of moderate-intensity exercise). A moderate effort is an effort leading to a noticeable increase in heart rate and respiratory rate or at the level of 11–13 points as measured by the Borg 20-degree rating of subjective exertion (RPE) (light-to-moderate effort) (Borg 1974). During a moderate effort, oxygen consumption increases three to six times as compared to resting oxygen consumption. During intensive exercise, oxygen consumption increases at least six times in comparison with resting oxygen consumption. The easiest way to control exercise intensity is measuring respiratory rate which should increase at least to a level allowing talking, but not singing.

Additionally, muscle-strengthening training involving exercises for 8–10 major muscle groups (8–15 repetitions of each exercise) should be done at least twice a week. Resistance exercise should begin with moderate intensity (level of 5–6 in a 10-point scale) and progress to high intensity (level 7–8). The resistance during exercise can be dosed with the aid of classic accessories used in resistance training (free weights, dumbbells, elastic bands) or everyday objects (a bottle with water, a can, a book). Resistance training is generally a safe form of activity, also for older individuals. However, to minimize the risk of injury and overloads, certain rules should be observed during resistance exercise performance, which are given as follows:

1. It is necessary to perform a warm-up before the basic training in order to prepare the circulatory and locomotor systems for an increased effort. The warm-up may take the form of simple, rhythmic workout performed without load, engaging all muscle groups subjected to resistance exercises.
2. It is advised not to start workout exercises with a heavy load. The recommended load level must be gradually achieved within a few training sessions.
3. It is necessary to avoid air retention during exercise. The workout should be coordinated with breathing. It is recommended to exhale while lifting up the load, and inhale when lowering the weight.
4. Movement should be executed smoothly and to the full extent.
5. During exercise, one should maintain a correct position, follow the principles of ergonomics and protect the spine against overload.
6. In the event of symptoms such as pain, shortness of breath, palpitations, dizziness or nausea, the exercise should be stopped, and in the case of persistence or repetition of the symptoms, it is necessary to consult a physician.
7. Persons with joint or periarticular tissue pain should not exercise.
8. Persons with chronic conditions should consult a physician to rule out possible contraindications for training.
9. Blood pressure and heart rate should be controlled periodically. Especially during the first training session, it is necessary to control blood pressure and heart rate to determine the cardiovascular response to effort.

The aforementioned program should be complemented by balance and flexibility exercises. The aim of stretching exercises is to maintain or increase the range of motion in the joints (Chodzko-Zajko et al. 2009). These exercises should be performed a minimum of two to three times per week. The stretching time should range from 10 to 60 seconds rather as static stretching, without violent, ballistic movements. The program should be enriched with balance exercises as they improve balance, especially in older people with impaired mobility and those prone to falls. There are no specific recommendations regarding the type, frequency and intensity of this kind of exercise. The proposed exercises include exercises gradually reducing the support plane, exercises on unstable surfaces, dynamic exercises that change the centre of gravity (e.g. changing the position by climbing on the toes or standing on the heels) and exercises performed with no visual stimuli (e.g. with eyes closed). Activities such as dancing (regardless of the style), or Tai-Chi or exercises on balance platforms also favourably affect one's balance and reduce the risk of falling.

It is recommended to gradually reach this level of PA. If older adults are not able to do the recommended amounts of exercise, they should be as physically active as possible. Exercise prescription should take into account ecological conditions of a target population (Rowinski et al. 2015). It has been shown that recommended levels of PA reduce all-cause mortality by 30%, and in active individuals, even with lower than recommended levels of PA, the risk of death is 20% than in physically inactive persons (Arem et al. 2015). Some researchers hypothesized that vigorous activity might bring even greater health benefits than moderate activities alone (Gebel et al. 2015). High-intensity interval training (HIIT) has recently started to be applied even in patients with cardiometabolic diseases (Weston et al. 2014). However, older people planning to change their lifestyles from sedentary to active should gradually reach the desired level of training intensity.

NUTRITION FOR OLDER PEOPLE

Nutrition and the quality of food as well as the exposure to harmful factors, smoking and drinking alcohol are important health determinants in the elderly population. The aspects relating to the nutritional state and nutrition are among the most important problems in elderly care (Kostka and Bogus 2007).

In geriatrics, there are two fundamentally different problems associated with nutritional status:

1. The problem of overweight and obesity, usually affecting younger seniors (60–74 years), living in their homes.
2. Protein-energy malnutrition (PEM) occurring primarily in seniors in advanced age (over 80 years old), with multiple coexisting diseases, usually hospitalized patients or nursing home residents.

ASSESSMENT OF NUTRITIONAL STATUS

Numerous approaches are used to assess one's nutritional status and the prevalence of overweight/obesity or malnutrition.

Anthropometric methods are very often used for such assessment as they take little time, can be easily implemented and are relatively cheap. In clinical practice, body mass index (BMI) which is a measure of weight adjusted for height is a widely used approach, measuring one's nutritional status:

$$BMI = bodymass \ (kg) / [height \ (m)]^2$$

It should be added that BMI values do not reflect the percentage of body fat in older people. According to the existing reports, the recommended BMI for older individuals is higher than that specified in WHO guidelines for younger adults, where the BMI ranging from 20 to 25 kg/m² is considered to be optimal. These data slightly differ, and depending on the studied population, it turns out, however, that seniors with BMI values within adult overweight/obesity range (even at the level of 29 kg/m²) have much higher survival rates and their quality of life is much better as compared to seniors having low BMI values (<25 kg/m² and thus recommended for younger adults). Although very high BMI values are associated with an increased risk of death, low BMI values also increase such risk. Seniors with higher BMI values are at a lower risk of falls and fractures. The reason for these differences is the age-related decline in lean body mass (LBM) and the decrease in body height due to vertebral compression and thoracic kyphosis, resulting in an altered relationship between BMI and fat mass (FM) (Beck and Damkjaer 2008).

Waist circumference (WC) and waist-to-hip ratio (WHR) are reliable indicators of visceral fat. The values – WC > 80 cm and WHR > 0.85 in women, and WC > 94 cm and WHR > 0.9 in men – indicate the presence of abdominal obesity. The percentage of body fat may be measured using the skinfold thickness measurement method, described by Durnin and Womersley (1974). Skinfold measurements are taken at four sites (triceps, biceps, sub-scapula and supra-ileum), and the percentage of body fat is estimated according to sex and age. However, the anthropometric methods, although simple and cheap, often fail to provide reliable, high-quality information about the nutritional status of older people.

There are other, more accurate body composition measurement techniques, such as hydrostatic weighing or doubly labelled water method (DLW). Imaging techniques are also used for this purpose. These include computed tomography, magnetic resonance imaging and dual-energy X-ray absorptiometry (DXA). Some of the approaches presented above have certain limitations due to the long and complicated measurement techniques, the requirements concerning the patient, a non-portable equipment, high financial cost or significant radiation.

Bioelectrical impedance analysis (BIA) is an interesting portable method to nutritional status assessment using electrical resistance. It involves measurement of impedance (total electrical resistance of the body) and takes advantage of the fact that individual tissues, due to varying water content, conduct electric current in a different way. The measurement is made using a set of body surface electrodes (the number of electrodes can be different) connected to the computer. The method enables assessment of a number of parameters, including FM, LBM, fat-free mass (FFM), the amount of water (i.e. total body water or TBW), intracellular water (ICW), extracellular water (ECW), body cell mass (BCM) and other parameters. It is a reliable, fast, secure, non-invasive, low-cost and reproducible approach to the analysis of body composition. It can be used in patients of all ages (children, adults and older people), both men and women, in healthy subjects and patients with chronic diseases (diabetes, hypertension, obesity, malnutrition). It is also recommended by the EWGSOP (European Working Group on Sarcopenia in Older People) as the method used in clinical practice for muscle mass assessment.

Mini Nutritional Assessment (MNA) is the most commonly used questionnaire for nutritional status assessment in older adults (Guigoz et al. 1994). The full version consists of 18 elements, which are a combination of a screening test and tools for nutritional status assessment. It includes questions relating to the essential elements of nutritional status, such as food intake, loss of body weight, the ability to move around independently, the incidence of acute disease or stress, neurological problems, BMI, the amount of medication taken, and assessment of the arm and calf circumferences. The total score that could be obtained in MNA questionnaire is 30. Values above 23.5 indicate a satisfactory nutritional status, while a score equal to or lover than 23.5 suggests malnutrition.

OVERWEIGHT AND OBESITY IN OLDER ADULTS

The age-related increase in body fat is due to numerous pathophysiological changes in body composition of older adults. LBM decreases with muscle mass starting to decline between 20 and 30 years of age, while the amount of body fat increases up to the 60th–70th year of age. The decline in TEE is caused by a reduction in basal metabolism rate (BMR), a reduced thermic effect of food (TEF) and lower levels of PA.

Obesity in adults increases the risk of pathological changes and diseases at any age, including all types of malignancies, arterial hypertension, atherosclerosis, coronary heart disease (CHD), myocardial infarction, stroke, respiratory diseases, degenerative changes of the musculoskeletal system, gout, fatty liver disease, gallstones, pancreatitis, diabetes, dyslipidaemia, depression and neurosis (Calle et al. 1999). In older adults, obesity is an important cause of frailty syndrome, decreased exercise tolerance and decreased cognitive functioning, the risk of falls, the risk of complications such as loss of autonomy, hospitalizations, institutionalization and even death (Wilson et al. 2002).

Any aggressive obesity treatment that can help younger adults is controversial in older age. Basic obesity treatment should combine moderate diet restriction with increased PA. Lifestyle changes are equally effective in older and younger individuals. It is particularly important to provide full coverage of the demand for protein, vitamins and minerals. The perfect and safe caloric deficit resulting from the combination of diet and PA should range from 200 to 500 kcal/day. A nutrient-rich diet coupled with PA should protect and increase muscle mass and LBM in senior citizens. A regular daily PA is a major determinant of the functional status and a preventive measure for disability in obese seniors. It has been shown that obesity is two to three times less common among seniors participating in regular, normal PA, while older people with high levels of PA are characterized by lower levels of body fat and higher muscle mass, which significantly improves their physical efficiency.

Protein-Energy Malnutrition in Older Adults

Malnutrition in old age, due to its high prevalence among older population as well as its complex causes and adverse implications, belongs to the so-called Geriatric Giants. The prevalence of PEM concerns mainly people in very advanced age, suffering from numerous chronic diseases, residing in nursing homes or staying in hospitals (Kostka et al. 2014a). Some studies have revealed a high incidence of malnutrition among institutionalized older patients (even from 20% to 78%). This phenomenon is associated with the between-group differences in the studied samples of patients and the different criteria of nutritional status assessment.

Inappropriate nutrient intake is associated with functional status decline and inferior quality of life in older adults. There are numerous factors contributing to the increased prevalence of malnutrition in older adults. These include age-related changes in body composition and the changes due to reductions in circulating anabolic hormones, malabsorption (celiac disease, pancreatic insufficiency), catabolic illness (chronic and systemic inflammatory diseases, tumours, infections) and other conditions, as well as stressful situations (thyroid gland diseases, depression, dementia, hospitalization, surgery), loss of appetite, reduction in the amount of saliva, dry mouth, dysphagia, hyposmia, quickly attained feeling of satiety, bad teeth/the need of getting dentures, medication use, disability, loneliness, poverty and difficulties in performance of such daily activities as doing shopping and preparing meals (Wysokiński et al. 2015).

There are three factors responsible for malnutrition: insufficient nutrient supply, excessive loss of vital nutrients following digestive disorders or malabsorption, and increased metabolic rate (hypermetabolism). Inadequate protein supply is the major cause of malnutrition, sarcopenia and disability (Cawood et al. 2012). Due to the decrease in muscle mass and slowing down of protein metabolism, the recommended standard consumption in seniors is higher as compared to that recommended for younger individuals. According to WHO, the daily demand for protein in adults is 0.91 ± 0.043 g/kg body mass/day, and the recommended dietary allowance (RDA) for protein in the United States equals 0.8 g/kg/day (Campbell et al. 1994). In older adults, the recommended daily protein intake is higher, amounting to 1.25 (1.0–1.5) g/kg/day, while the recommended supply of protein that plays a role in the prevention of sarcopenia is 1.2–1.5 g/kg/day (Deutz et al. 2014).

The data, however, indicate that the calorie intake among older adults is often inadequate and accompanied by insufficient supply of vitamins and minerals in their diets. Excessive energy supply resulting from high consumption of fat and carbohydrates may lead to metabolic disorders (Guligowska et al. 2015). Moreover, it is important for older adults to observe the recommended dietary supplementation with calcium and vitamin D, with the latter playing a role in prevention and treatment of osteoporosis. Finally, it is recommended to intake at least eight glasses of water/fluids a day (30–35 ml/kg body weight). This is extremely important, because older adults are particularly vulnerable to the risk of dehydration. In addition, adequate amounts of fluid protect the kidneys and prevent infections of the urinary tract, the formation of kidney stones and constipation.

REHABILITATION IN OLDER INDIVIDUALS INCLUDING REHABILITATION TO RESTORE WORK ABILITY

Ageing of populations is a phenomenon that mankind has never experienced before. The age-associated decline in physical capacity and the increasing rates of disability in developed societies may well prove to be the most significant challenges faced by public health in the coming years.

Human ageing is associated with a gradual decrease in functional capacity and the presence of numerous chronic diseases. Such changes result in the increased percentage of people with disabilities. Maintaining and improving their functional status through rehabilitation is an important challenge of geriatric care. Rehabilitation is a complex process aimed at restoring patient independence as far as possible, enabling the patients to function independently within the family and the community, participating in social activities and, sometimes, regaining work ability. Medical rehabilitation is an essential part of the wider rehabilitation process, aimed at restoring numerous body functions after disease or trauma.

There are some differences in rehabilitation procedures applied in young people and older adults. While young people tend to fully restore their lost skills after disease or accident as well as their social status, it often proves to be unsuccessful in older age. Due to the prevalence of comorbidities, a lower potential for restoring the lost fitness levels as well as the financial and social barriers, the rehabilitation of older people is not an easy process. A number of such people require rehabilitation proceedings not only because of illness, but also due to their age-related decrease in functional efficiency. Therefore, rehabilitation programs for seniors should be mainly focused on disability prevention.

The main target of geriatric rehabilitation is not to prevent death, but to overcome functional decline and disability in later life. An increasing body of scientific research regarding functional capacity concerns older people and their declining level of functional status with advancing age (Rolland et al. 2008). Physical disability due to ageing, prevalent diseases and an inactive lifestyle can lead to a loss of ability in performing the basic activities of daily living and consequently, to the loss of independence in daily life. Lack of independence in such activities is a predictor of falls (Delbaere et al. 2010), institutionalization (Abellan van Kan et al. 2009), hospitalization (Studenski et al. 2003), deterioration of the quality of life and increased mortality (Blain et al. 2010). Therefore, it is very important to identify the factors determining the functional performance of older patients and to develop preventive and rehabilitation programs designed to improve the dimensions that have the greatest impact on their functional health.

The rehabilitation of older people must take into account the hierarchy of patient needs, focusing primarily on maintaining and restoring the ability to perform basic (ADL) and then complex (IADL) activities of everyday life.

Despite its complexity, rehabilitation of older people rarely has to be carried out in specialized centres. Depending on the circumstances and the patient's health status, needs and abilities, it can be performed in hospital, nursing home or clinic settings, and even at the patient's home. As far as possible, the rehabilitation team should be composed of doctors of various specialties, physical therapists, nurses, social workers, a speech therapist and a psychologist. A specialist in rehabilitation or a geriatrician should be responsible for coordination of their activities (Szybalska et al. 2018).

There are numerous cases requiring individual rehabilitation programs, but it is often advantageous to conduct group exercises (including contact with the group, meeting new persons, additional mobilization and motivation for exercises). Due to the multitude of age-related problems, the process of rehabilitation must be multidimensional; however, it should be targeted at the current critical health problem(s), e.g. stroke, heart attack, pneumonia or fracture of the femoral neck.

Medical rehabilitation is a fundamental part of the entire rehabilitation process, which includes therapeutic procedures (kinesitherapy, using electrophysical agents and massage), occupational therapy, pharmacotherapy, psychotherapy and psychological counselling, and orthopaedic supplies. Social and occupational rehabilitations are also important components of rehabilitation programs, and the latter, however, is less often applied in older patients (Sołtysik et al. 2017).

Rehabilitation procedures should be individually planned and always focused on functional efficiency. Depending on the current condition of the patient and his potential for rehabilitation, such procedures should focus on (1) prevention of function loss, (2) slowing down the pace of function loss, (3) improvement and/or the restoration of function, (4) compensation for the lost function and (5) maintaining the function.

The World Health Organization has developed recommendations for all healthy older people, older people with chronic diseases and disabilities, and seniors with a variety of ailments, who may need to take additional precautions (medical consultation, additional diagnosis) and modification of the program. The recommendations pertain to participation in PA and include the following: Aerobic (endurance) exercise: moderate intensity: at least 150 minutes/week (e.g. 5 times a week for 30 minutes) or intense: at least 75 minutes/week (e.g. 5 times a week for 15 minutes) or a combination of moderate and high output (e.g. 3 times a week for 30 minutes moderate effort and 2 times a week after 15 minutes. intensive effort).

For larger benefits, the time of training should be increased to 300 minutes of moderate activity or 150 minutes of activity a week. A single exercise session should not last less than 10 minutes. Resistance exercises should be done at least two times per week and involve the most important muscle groups. Exercises to improve balance and reduce the risk of falls should be performed at least three times a week.

REHABILITATION AFTER CANCER

The number of cancer survivors increases rapidly owing to early detection of neoplastic changes and the advances in treatment. Sustaining physical and mental abilities and returning to work are the major objectives of care provided for such patients. Therefore, oncology specialists will have to coordinate patient care with cardiologists, geriatricians and primary care providers. As compared to the general population, cancer survivors have usually been reported to be at higher risk of late complications and development of a variety of concomitant diseases, including cardiovascular disorders. Cardiovascular changes include organ damage and increased CVD risk after radiotherapy, chemotherapy, hormonal therapy and targeted therapies (Schmitz et al. 2010). Identification and reduction of these late effects will become increasingly important. Management of comorbid conditions and disability prevention should be a standard of care for older cancer survivors.

Patients after cancer are at risk for complications, other diseases and decline in overall PA.

Historically, cancer patients were usually advised to avoid PA. Nowadays, the growing population of cancer survivors and the emerging research outcome have challenged the recommendations related to PA (Schmitz et al. 2010).

It seems that increased PA is beneficial and safe in the majority of cancer patients.

Physical exercise has been shown to improve fitness levels, physical functioning and the quality of life in cancer patients (McTiernan 2004). In the general population, PA reduces risk for several types of cancer, but the effects of PA on survival rate have not been so unequivocally documented so far in cancer patients.

A roundtable session convened by the American College of Sports Medicine (ACSM) has concluded that exercise training is safe during and after cancer treatment, and results in improvements in physical functioning, the quality of life and reduction of cancer-related fatigue in several cancer survivor groups (Schmitz et al. 2010). ACSM recommends that cancer survivors should "avoid inactivity" and follow the 2008 Physical Activity Guidelines for Americans, with specific exercise programming adaptations (Schmitz et al. 2010). In the follow-up paper, a process for implementing the guidelines in clinical practice was outlined, and recommendations were provided for oncology care providers on how they can interface with the exercise and physical therapy community (Wolin et al. 2012).

The meta-analysis of 71 cohort studies has shown that the hazard ratio (HR) values obtained from the individuals who participated in the highest number of physical activities were 0.83 (95% CI 0.79–0.87) and 0.78 (95% CI 0.74–0.84) compared with the values corresponding to cancer mortality in the general population and among cancer survivors, respectively. There was an inverse non-linear dose-response between the effects of PA and mortality rate in cancer patients. In the general population, a minimum of 2.5 hours/week of moderate-intensity activity resulted in a significant 13% decrease in cancer mortality. The risk of death due to cancer was 27% lower in cancer survivors who completed 15 metabolic equivalents of task (MET)-hours/week of PA. A greater protective effect was noted in cancer survivors undertaking PA postdiagnosis versus prediagnosis, with 15 MET-hours/week contributing to decreased mortality risk by 35% and 21%, respectively. The results of the meta-analysis support the statement that current WHO PA recommendations concerning PA have contributed to reduction of cancer-related mortality in both the general population and cancer survivors. The authors conclude that PA in patients diagnosed with cancer has a protective effect, preventing recurrence of the disease (Li et al. 2016).

The guidelines issued by 2007 World Cancer Research Fund/American Institute for Cancer Research (WCRF/AICR), concerning diet and PA, encourage cancer survivors to follow cancer prevention recommendations. In 2193 older women with a confirmed cancer diagnosis, adherence to the WCRF/AICR recommendations was associated with the improvement of health-related quality of life (HRQOL). According to Inoue-Choi et al. (2013), "Following the lifestyle guidelines for cancer prevention may lead to HRQOL improvement among elderly female cancer survivors. PA may be a key lifestyle factor to improve HRQOL".

Finally, the results of research on the benefits of PA in cancer patients suggest that PA interventions are safe for cancer survivors and improve their fitness levels, strength, physical function and cancer-related psychosocial variables, whereas dietary interventions improve diet quality, nutrition-related biomarkers and body weight. Pekmezi and Demark-Wahnefried (2011) maintain that "Preliminary evidence also suggests that diet and exercise may positively influence biomarkers associated with progressive disease and overall survival (e.g. insulin levels, oxidative DNA damage, tumor proliferation rates)". Although there are no doubts that more research is needed to address specific types of cancer and approaches to cancer treatment, the general recommendation for "avoiding inactivity" should be followed by cancer survivors.

In cancer survivors, return-to-work (RTW) has become one of the key issues in preventive and curative treatment. Accumulating evidence suggests that awareness of patients and health professionals as well as appropriate employment and social policies may allow efficacious interventions for RTW (Kiasuwa Mbengi et al. 2018).

The recent data indicates that lower levels of fatigue as well as a higher value of work, work ability and job self-efficacy are associated with earlier RTW in cancer patients after chemotherapy. The authors point out that occupational rehabilitation should motivate patients to create a positive and safe working environment and to educate them on rights and obligations during sick leave (Wolvers et al. 2018).

In the study conducted in 837 employed prostate cancer survivors, the RTW rate was 62% at 12-month follow-up, while the survivors with lower socioeconomic status showed least favourable outcomes (Ullrich et al. 2017).

Kiasuwa Mbengi studied 15,543 Belgian cancer survivors qualified for work disability. The overall median time of work disability in the studied sample was 1.59 years, ranging from 0.75 to 4.98 years. The factors indicating their ability to RTW included the following: "Being women, white collar, young and having haematological, male genital or breast cancers" were factors contributing to the best likelihood of being able to RTW (Kiasuwa Mbengi et al. 2018).

In a systematic review of cohort and case–control studies conducted in colorectal cancer patients, the factors having a significantly negative impact on their RTW included: (neo) adjuvant therapy, age (they were older) and more comorbidities (den Bakker et al. 2018). The factors increasing the risk of work disability included previous period of unemployment, extensive surgical resection and postoperative complications work disability. The authors conclude that healthcare professionals need to be aware of these prognostic factors to select patients eligible for timely intensified rehabilitation in order to optimize the RTW process and prevent work disability (den Bakker et al. 2018).

In the Netherlands, a multidisciplinary rehabilitation programme was developed for cancer survivors who planned to return to work. The program combined occupational counselling with a supervised physical exercise programme during chemotherapy. After the start of a multidisciplinary rehabilitation programme that combined occupational counselling with a supervised physical exercise programme, 59% of the cancer patients returned to work within 6 months, while 86% and 83% returned to work within 12 and 18 months, respectively. In addition, significant improvements ($p < 0.05$) were noted in the patients' attitudes towards work, work ability, self-efficacy after RTW and the quality of life, whereas their fatigue levels were significantly reduced (Leensen et al. 2017).

REHABILITATION IN CARDIOVASCULAR AND METABOLIC DISEASES

Physical and mental health status, concurrent diseases and social environment of older adults with cardiovascular diseases are the factors that should be considered in personal recommendations for this population, concerning the role of exercise in CVD prevention. In older adults, CVD usually coexist with other comorbidities and functional problems such as sarcopenia, frailty or malnutrition. Exercise-based cardiac rehabilitation has been shown to reduce the risk of cardiovascular mortality and the risk of hospitalization, and to improve quality of life (Anderson et al. 2016). Such benefits may be less clear in multimorbid older adults, e.g. in diabetic patients with long-term mortality increased by 50% following acute myocardial infarction, as compared to patients without such health problems (Gholap et al. 2016). Nevertheless, the overall current evidence indicates that involvement in PA or even sporting activities is beneficial and generally safe, for both healthy older adults and multimorbid or frail ones, provided that adequate safety issues are taken into consideration.

The protective effect of PA against CVD is due to, among others, a favourable modification of changes in the circulatory system, which are usually associated with age (Lemura et al. 2000). Regular moderate and/or vigorous exercise increases aerobic capacity (VO_2max) and delays its age-related decline. Aerobic capacity diminishes with age by about 10% per decade. Therefore, even a small increase in VO_2max may extend the period of functional independence by several years. In older people, aerobic performance can be improved by both aerobic exercise and resistance training. This is due to the fact that oxygen uptake efficiency in the elderly population depends to a greater extent on peripheral mechanisms (muscle oxidative capacity and the ability to use oxygen) than in young people. Therefore, resistance training, resulting in muscle mass and function improvement, also contributes to aerobic capacity improvement (Maiorana et al. 2000).

Physical exercise has been advocated for decades in prevention and treatment of overweight/obesity, diabetes and other metabolic disorders. Physical training affects a number of so-called modifiable risk factors for CVD, including beneficial effect on lipid profile (increase in HDL cholesterol, lower LDL and triglycerides), glucose tolerance improvement and prevention of overweight and obesity. Inadequate exercise (<1.07 METh/d run or walked) is a risk factor for sepsis-associated mortality, particularly in diabetic patients (Williams 2013).

The updated Cochrane review supports the conclusions that, compared with no exercise control, exercise-based cardiac rehabilitation reduces the risk of cardiovascular mortality, but it doesn't reduce total mortality (Anderson et al. 2016). Cardiac rehabilitation is associated with reduced mortality even in the modern era of treatment with statins and acute revascularization (Rauch et al. 2016). Besides, it was found that cardiac rehabilitation significantly reduced the risk of hospitalization, but not in patients at risk of myocardial infarction or after revascularization. Exercise-based cardiac rehabilitation was also found to improve HRQL (Anderson et al. 2016).

Running or walking decreases CVD mortality risk progressively at most exercise levels in patients after a cardiac event, but the role of exercise in CVD mortality prevention is attenuated at the highest levels of exercise (running: above 7.1 km/d or walking briskly: 10.7 km/d) (Williams and Thompson 2014).

RTW is one of the major problems in patients after acute coronary events. People with CHD often require prolonged absences from work to convalescence after acute disease events, such as myocardial infarctions or after revascularization procedures. A reduced functional capacity and anxiety due to CHD may further delay or prevent RTW (Hegewald et al. 2019).

Using individual-level linkage of Danish Nationwide Registry data, 39,296 patients of working age who were employed before admission and discharged after the first-time myocardial infarction in the period between 1997 and 2012 were identified (Smedegaard et al. 2017). Despite the fact that most patients returned to work, about one in four was detached from employment after 1 year. Several factors including age and a lower socioeconomic status were associated with risk of detachment from employment (Smedegaard et al. 2017).

The data on the RTW after the first hospitalization for heart failure in Denmark from 1997 to 2012 shows that patients more likely to RTW were younger men, with higher education and income, without a history of stroke, diabetes, chronic kidney disease, chronic obstructive pulmonary disease (COPD) or cancer (Rørth et al. 2016).

The recent review of randomized controlled trials (RCTs) examined RTW among people with CHD who were provided either person-directed, psychological counselling interventions or usual care (Hegewald et al. 2019). These interventions did not increase RTW up to 6 months or at 6 to 12 months. These psychological interventions did not shorten the time until RTW. The authors concluded that psychological interventions may have little or no effect on the proportion of participants working between one and five years. Combined cardiac rehabilitation programmes increased RTW up to 6 months, and probably not at 6 to 12 months' follow-up. Combined interventions increased RTW up to 6 months and probably reduced the time away from work. Otherwise, the authors found no evidence of either a beneficial or harmful effect of person-directed interventions (Hegewald et al. 2019).

Salzwedel et al. (2019) emphasize the importance of using a multimodal approach to cardiac rehabilitation, in particular, its psychosocial components. In a sample of 401 patients below 65 years of age, participating in the 3-week inpatient cardiac rehabilitation, perception and expectation regarding the occupational prognosis played a pivotal role in predicting their return to work 6 months after acute coronary syndrome (Salzwedel et al. 2019).

REHABILITATION IN RESPIRATORY DISEASES

Pulmonary rehabilitation is widely used in patients with respiratory system diseases. Pulmonary rehabilitation may be applied in patients with all types of acute and chronic lung disease, especially in COPD, and in those before and after lung surgery. Rehabilitation for patients with pulmonary diseases improves their functional capacity and quality of life, and there are data indicating reduction in the number of hospitalizations and mortality rates after such procedures.

Pulmonary rehabilitation includes exercise, education and lifestyle modification programs. Physical training usually includes endurance and resistance exercises. Exercise training programme seems effective in improving exercise capacity and HRQOL, which is confirmed by short-term and 6-month follow-up studies (Dale et al. 2015).

The review of Cochrane Database of Systematic Reviews has shown that pulmonary rehabilitation relieves dyspnoea and fatigue, improves emotional function and enhances the sense of control that individuals have over their condition. The above-mentioned improvements are moderate, yet clinically significant (McCarthy et al. 2015). There is no important data in the scientific literature on RTW in patients with respiratory diseases undergoing pulmonary rehabilitation.

REHABILITATION IN MUSCULOSKELETAL DISEASES

PA, together with dietary intervention, is a cornerstone in prevention and treatment of sarcopenia, frailty and malnutrition (Beaudart et al. 2016). Musculoskeletal disorders and CVD are most prevalent in older adults. PA, especially progressive resistance and balance training with calcium and vitamin D supplementation, has long been recommended for the prevention and treatment of osteoporosis. Nevertheless, minimizing the risk of falls and fractures by adherence to safety guidelines, e.g. avoiding exercises with loaded spine flexion, is essential (Beck et al. 2017). Long-term aerobic and muscle strengthening exercise of moderate to high intensity reduces activity limitations and improves both oxygen uptake and muscle strength in adults with rheumatoid arthritis (RA) (Swärdh and Brodin 2016). PA is recommended for relief of pain and functional status improvement in patients with osteoarthritis of the knee or hip joints. It has been suggested that greater health benefits may be attained with high-intensity rather than low-intensity exercise or PA (Regnaux et al. 2015). Nevertheless, the recent review found low-quality evidence for clinical benefits of high-intensity compared with low-intensity exercise programs including such short-term effects as pain relief and physical function improvement (Regnaux et al. 2015).

The functional state of skeletal muscles is one of the most important predictors of functional capacity in advanced age; however, the varied properties of muscles may be due to different underlying mechanisms. So far, multiple factors affecting muscle strength and power have been described. These include age, genetic factors, muscle cross-section, gender, malnutrition, health status and associated diseases, behavioural factors, such as levels of PA, smoking, hormonal factors, the concentration of vitamin D, the concentration of cytokines and even race (Norman et al. 2010). Health status and concomitant diseases are associated with the quality of muscle work. Concomitant diseases contribute to the limitation of PA and, in consequence, to further loss of muscle mass. The impact of cardiovascular diseases on muscle function is associated with lower blood flow, hypoxia and oxidative stress. The reduction of muscle mass after stroke is largely dependent on the level of PA reduction. Muscle weakness and loss of strength concern mainly the affected limbs, but they apply to both sides (English et al. 2010). The pain due to the presence of osteoarthritis is the primary cause of the decline in PA and the associated decrease of muscle mass (Cecchi et al. 2008). On the other hand, the reduction in muscle strength is considered as a risk factor for degenerative changes. In patients with osteoarthritis, thigh muscle strength was found to be 16%–27% lower as compared to their peers without such symptoms (Hinman et al. 2010). Lower leg muscle power correlates with the presence of pain in the hip and knee (Cecchi et al. 2008). Finally, pharmacotherapy may also influence muscle function. For example, long-term corticosteroid intake in patients with rheumatic diseases was associated with atrophy of type II fibres and a reduced capacity to generate muscle power (Rothstein et al. 1983).

Sustaining a satisfactory PA level and nutritional status combined with effective treatment of accompanying diseases are the cornerstones of proper neuromuscular function in advanced age. PA is the best solution contributing to improvement of muscle strength and power. The most effective means of training, aimed at muscle strength increase is resistance training, aerobic training, however, also enhances muscle, affects the neural mechanisms responsible for proper functioning of the muscles, stimulates protein synthesis and reduces body fat (Rolland et al. 2008). Thanks to these potential benefits, the ACSM recommends the inclusion of resistance training as part of an exercise program for older adults (Chodzko-Zajko et al. 2009). An increase in muscle strength without a significant increase of muscle mass is attributed to the improved recruitment of motor units by the nervous mechanisms

(Gabriel et al. 2006). The improvement of muscle power is mainly due to increases in muscle strength (Porter 2006). Increases in the velocity of muscle shortening can probably be explained by improved capacity to recruit fast muscle fibres.

Future studies will concentrate on the efficacy and feasibility of power- and velocity-oriented training in patients with different diseases and different functional status. The training protocols in some studies that use higher-velocity training suggest that the gains in power in such training may be greater than those in classic resistance training (Chodzko-Zajko et al. 2009). Some benefits of speed training have already been described in high-functioning older adults and in patients with early Parkinson's disease (de Vos et al. 2008, Pohl et al. 2003). The optimization of the metabolic and hormonal environment for muscle power generation will also be an important challenge for future studies (Kostka et al. 2000).

Regular PA in old age is widely recommended as an effective way to prevent chronic diseases and to maintain well-being. The view that PA has a beneficial effect on the health status, functional performance and seniors' quality of life is widely accepted. PA plays an important role in the prevention and management of a number of conditions such as CHD, stroke, diabetes, osteoporosis, some types of cancer or mental health disorders. Older people generally undertake lower-intensity activities (gardening, walking, golf, low-intensity exercises) than younger people (undertaking running or high-intensity exercises). On the other hand, increased participation in sport has been consistently observed among the older subject. Participation in sports carries the risk of falls and injuries. In older people, the risk of injury is greater due to age-related pathophysiological changes and concomitant chronic conditions. Available data indicate an increasing number of injuries among the elderly, which is associated with both ageing of the population and an increasing number of older people practicing sports and physical exercises. An appropriate identification of risk factors for injury and education of older people can reduce the incidence of injuries. Injury prevention strategies include using helmets and protective equipment, warming up and a properly designed training program. The health benefits of participation in regular PA adjusted to health status and physical functioning outweigh hazards of sport-related injuries, even in advanced age.

There are no specific guidelines restricting participation in even the most intense forms of exercise for older individuals or indicating when to stop practicing sports. The American College of Sports Medicine/American Heart Association (ACSM/AHA) include vigorous-intensity activities in their recommendations on PA for older adults and recommend only "Any modality that does not impose excessive orthopedic stress". Moreover, the Department of Health and Human Services (DHHS), in their PA Guidelines for Americans note that additional health benefits occur with the increase in the amount of PA of longer duration, performed with higher intensity and greater frequency (Chodzko-Zajko et al. 2009).

On the other hand, a low level of PA entails a number of consequences that increase the risk of falls and thus the risk of injury. In older people, the risk of injury is also higher due to age-related changes and concomitant conditions. Thus, although participation in regular PA should be promoted, attention should also be focused on another health goal: to decrease the incidence of injuries associated with sport and recreational activities.

Osteoarthritis is probably one of the most important diseases hampering regular exercises because of chronic pain (American Geriatrics Society Panel on Exercise and Osteoarthritis 2001). Nevertheless, moderate-intensity multimodal rehabilitation together with appropriate pharmacotherapy and surgical interventions in severe cases should be the treatment of choice in attenuating further functional decline and maintaining satisfactory functional abilities (Kostka et al. 2014b).

Besides pharmacological treatment, regular exercise is one of the cornerstones of care in RA. In many of the earlier studies of exercise in RA, the intensity exercise did not reach the currently recommended level or is not described in satisfactory detail. A recent narrative review indicates that there is moderate-quality evidence that short-term land-based aerobic exercise of moderate to high-intensity augments oxygen uptake but does not improve muscle strength. Short-term water-based aerobic exercise of moderate- to high-intensity augments oxygen uptake; short-term land-based aerobic and muscle strengthening exercise of moderate- to high-intensity augments oxygen uptake and muscle strength. Long-term land-based aerobic and muscle strengthening exercise of moderate to high intensity reduces activity limitations and improves both oxygen uptake and muscle strength. Clinicians should recommend that patients with RA to participate in various types of exercise (Swärdh and Brodin 2016).

REHABILITATION IN NERVOUS SYSTEM AND PSYCHIATRIC DISEASES

Neurodegenerative disorders constitute a growing problem in older populations. There is some promising evidence that exercise programs may improve the functional abilities in people with dementia and mild cognitive impairment (Forbes et al. 2015). Active older people achieve better results in tests assessing cognitive function compared with their inactive peers. It is estimated that about 13% (nearly 4.3 million) cases of Alzheimer's disease worldwide can be caused by the lack of PA. A 10% reduction of the prevalence of physical inactivity can potentially allow to avoid/delay more than 380,000 cases of Alzheimer's disease, whereas a 25% reduction of inactivity can reduce the prevalence of this disease in the world by almost 1 million cases. The impact of physical exercise on the maintenance of cognitive functions is explained by several possible mechanisms, such as an increase in cerebral blood flow, an increase in cerebral metabolism, reduction of risk factors for cardiovascular disease, stimulation of neuron growth and survival. PA also promotes social activities and integration, which can have a positive effect on cognitive functioning. Regular PA seems equally effective in improving physical and cognitive functional capacities in patients with Parkinson's disease (Lauzé et al. 2016). On the other hand, PA seems less efficient in terms of improvement of clinical symptoms in patients with Parkinson's disease and psychosocial aspects of life, with only 50% or less of the results showing positive effects. The impact of PA on cognitive functions and depression also appears weaker, but only few studies have examined these outcomes (Lauzé et al. 2016).

In Parkinson's disease and Alzheimer's disease, physical therapy is generally oriented on movement disorders as consequences of the disease. It can, however, improve the functional state of patients, increasing their independence level and improving their quality of life. Exercise should be tailored to the stage of deficits and the patient's needs, and focused on daily life tasks.

Acquired brain injury is known to be severely disabling. On average, 40% of employees RTW within two years after injury. Women and patients with non-comorbid impairments returned to work earlier than men and patients with multiple impairments (Aas et al. 2018).

Despite advances in stroke therapy, many patients still face cognitive, emotional and physical impairments (Kostka et al. 2019). Stroke is a leading cause of serious long-term disability and subsequent failure to RTW. According to the literature review, one of the most consistent predictors of RTW was stroke severity. The patients who experienced a mild to moderate stroke, those of Caucasian ethnicity and those with a higher socioeconomic status were more likely to RTW (Ashley et al. 2019). Rehabilitation is here the most important element of the procedure, and any contraindications to it (unstable clinical condition, inflammatory diseases, lack of motivation on the part of the patient) should be treated as far as possible as temporary. The basic purposes of rehabilitation in post-stroke patients include reduction of mortality, prevention of complications, induction of compensating mechanisms, reduction of the degree of disability, promotion of functional independence and improvement of the quality of life in this group of patients. Cognitive impairments are highly prevalent in stroke survivors and can substantially affect their physical rehabilitation outcome and their quality of life. The management of these impairments currently remains limited, but the increasing number of studies report the effect of aerobic exercise on cognitive performance in patients suffering from stroke. Available evidence indicates that aerobic exercise may have a positive effect on global cognitive ability improvement and a potential beneficial effect on memory, attention, the visuospatial domain of cognition and the quality of life in stroke survivors (Zheng et al. 2016). Rehabilitation after stroke is especially difficult in older people due to the limited rehabilitation potential and a lesser ability to improve their neurological conditions compared to younger patients. In the rehabilitation proceedings, realistic goals and the deadlines for achieving them must be set.

Chronic fatigue syndrome (CFS) is characterized by persistent, medically unexplained fatigue, as well as such symptoms as musculoskeletal pain, sleep disturbances, headaches and impaired concentration, and short-term memory. CFS presents as a common, debilitating and serious health problem. Patients with CFS may generally benefit from exercise therapy and feel less fatigued following it. There is no evidence suggesting that exercise therapy may worsen health-related outcomes. A positive effect of exercise therapy on sleep, physical function and self-perceived general health has been observed, "but no conclusions for the outcomes of pain, quality of life, anxiety, depression, drop-out rate and health service resources were possible". (Larun et al. 2016).

PREHABILITATION IN OLDER ADULTS (PREPARATION FOR SURGERY)

Pre-operative assessment of older patients, especially those with frailty syndrome, allows the prediction of operational outcomes, potential complications, risk of death as well as the length of the patient's stay in the hospital after surgery and possible needs for care services including nursing home (Robinson et al. 2015).

The study conducted by Makary et al. (2010) in patients diagnosed with frailty syndrome before surgery revealed 2.5 times higher risk of postoperative complications (OR 2.54; 95% CI 1.12-5.77) and more than 20 times higher risk of institution care placement (OR 20.48; 95% CI 5.54-75.68) among people living in their home environment before the surgery.

The 2.4-year follow-up examination carried out in almost 30,000 Europeans aged above 50 years revealed that deficit accumulation frailty index was a stronger predictor of mortality than age (Romero-Ortuno and Kenny 2012).

There have been reports indicating that the programs carried out before the surgery (3–8 weeks), primarily based on exercise and dietary interventions, increase the functional reserve of patients and can improve the operating outcome and reduce the risk of complications (prehabilitation). Such programs include

1. Optimization of health status by controlling the coexisting diseases, discontinuing harmful habits such as smoking, and controlling and modifying the amount of medications taken.
2. PA/exercises, especially aerobic training and exercises, increasing muscle mass and muscle strength (resistance exercises), and breathing exercises.
3. Dietary interventions related to supplementation of nutritional deficiencies, as well as the optimization of the nutrients before the procedure to compensate for the catabolic processes after the treatment and prevent the loss of LBM.
4. Decreasing anxiety and mood perturbations.

Prehabilitation programme adapted and modified according to the patient's condition should be continued also during the postoperative period.

REFERENCES

Aas, R. W., L. A. Haveraaen, E. P. M. Brouwers, and L. S. Skarpaas. 2018. Who among patients with acquired brain injury returned to work after occupational rehabilitation? The rapid-return-to-work-cohort-study. *Disabil Rehabil* 40(21):2561–2570.
Abellan van Kan, G., Y. Rolland, S. Andrieu et al. 2009. Gait speed at usual pace as a predictor of adverse outcomes in community-dwelling older people an International Academy on Nutrition and Aging (IANA) Task Force. *J Nutr Health Aging* 13(10):881–889.
ACSM. 2009. American College of Sports Medicine position stand. Progression models in resistance training for healthy adults. *Med Sci Sports Exerc* 41(3):687–708.
American Geriatrics Society Panel on Exercise and Osteoarthritis. 2001. Exercise prescription for older adults with osteoarthritis pain: consensus practice recommendations. A supplement to the AGS Clinical Practice Guidelines on the management of chronic pain in older adults. *J Am Geriatr Soc* 49(6):808–823.
Anderson, L., D. R. Thompson, N. Oldridge et al. 2016, Jan. 5. Exercise-based cardiac rehabilitation for coronary heart disease. *Cochrane Database Syst Rev* (1):CD001800.
Arem, H., S. C. Moore, A. Patel et al. 2015. Leisure time physical activity and mortality: a detailed pooled analysis of the dose-response relationship. *JAMA Intern Med* 175(6):959–967.

Ashley, K. D., L. T. Lee, and K. Heaton. 2019. Return to work among stroke survivors. *Workplace Health Saf* 67(2):87–94.

Beaudart, C., E. McCloskey, O. Bruyère et al. 2016, Oct. 5. Sarcopenia in daily practice: assessment and management. *BMC Geriatr* 16(1): 170.

Beck, A. M., and K. Damkjaer. 2008. Optimal body mass index in a nursing home population. *J Nutr Health Aging* 12(9):675–677.

Beck, B. R., R. M. Daly, M. A. Singh, and D. R. Taaffe. 2017. Exercise and Sports Science Australia (ESSA) position statement on exercise prescription for the prevention and management of osteoporosis. *J Sci Med Sport* 20(5):438–445.

Blain, H., I. Carriere, N. Sourial et al. 2010. Balance and walking speed predict subsequent 8-year mortality independently of current and intermediate events in well-functioning women aged 75 years and older. *J Nutr Health Aging* 14(7):595–600.

Borg, G. A. 1974. Perceived exertion. *Exerc Sport Sci Rev* 2:131–153.

Calle, E. E., M. J. Thun, J. M. Petrelli, C. Rodriguez, and C. W. Jr Heath. 1999. Body-mass index and mortality in a prospective cohort of U.S. adults. *N Engl J Med* 341(15):1097–1105.

Campbell, W. W., M. C. Crim, G. E. Dallal, V. R. Young, and W. J. Evans. 1994. Increased protein requirements in elderly people: new data and retrospective reassessments. *Am J Clin Nutr* 60(4):501–509.

Cawood, A. L., M. Elia, and R. J. Stratton. 2012. Systematic review and meta-analysis of the effects of high protein oral nutritional supplements. *Ageing Res Rev* 11(2):278–296. DOI: 10.1016/j.arr.2011.12.008.

Cecchi, F., A. Mannoni, R. Molino-Lova et al. 2008. Epidemiology of hip and knee pain in a community based sample of Italian persons aged 65 and older. *Osteoarthr Cartilage* 16(9):1039–1046.

Chodzko-Zajko, W. J., D. N. Proctor, M. A. Fiatarone Singh et al. 2009. American College of Sports Medicine position stand. Exercise and physical activity for older adults. *Med Sci Sports Exerc* 41(7):1510 1530.

Dale, M. T., Z. J. McKeough, T. Troosters, P. Bye, and J. A. Alison. 2015, Nov. 5. Exercise training to improve exercise capacity and quality of life in people with non-malignant dust-related respiratory diseases. *Cochrane Database Syst Rev* (11):CD009385.

de Vos, N. J., N. A. Singh, D. A. Ross, T. M. Stavrinos, R. Orr, and M. A. Fiatarone Singh. 2008. Effect of power-training intensity on the contribution of force and velocity to peak power in older adults. *J Aging Phys Activity* 16(4):393–407.

Delbaere, K., J. C. Close, J. Heim et al. 2010. A multifactorial approach to understanding fall risk in older people. *J Am Geriatr Soc* 58(9):1679–1685.

den Bakker, C. M., J. R. Anema, A. G. N. M. Zaman et al. 2018. Prognostic factors for return to work and work disability among colorectal cancer survivors; A systematic review. *PLoS One* 13(8):e0200720.

Deutz, N. E., J. M. Bauer, R. Barazzoni et al. 2014. Protein intake and exercise for optimal muscle function with aging: recommendations from the ESPEN Expert Group. *Clin Nutr* 33(6):929–936. DOI: 10.1016/j.clnu.2014.04.007.

Durnin, J. V., and J. Womersley. 1974. Body fat assessed from total body density and its estimation from skinfold thickness: measurements on 481 men and women aged from 16 to 72 years. *Br J Nutr* 32:77–97.

English, C., H. McLennan, K. Thoirs, A. Coates, and J. Bernhardt. 2010. Loss of skeletal muscle mass after stroke: a systematic review. *Int J Stroke* 5(5):395–402.

Forbes, D., S. C. Forbes, C. M. Blake, E. J. Thiessen, and S. Forbes. 2015, Apr. 15. Exercise programs for people with dementia. *Cochrane Database Syst Rev* (4):CD006489. DOI: 10.1002/14651858.CD006489.pub4.

Gabriel, D. A., G. Kamen, and G. Frost. 2006. Neural adaptations to resistive exercise: mechanisms and recommendations for training practices. *Sports Med* 36(2):133–149.

Garber, C. E., B. Blissmer, M. R. Deschenes et al. 2011. American College of Sports Medicine position stand. Quantity and quality of exercise for developing and maintaining cardio-respiratory, musculoskeletal, and neuromotor fitness in apparently healthy adults: guidance for prescribing exercise. *Med Sci Sports Exerc* 43(7):1334–1359.

Gebel, K., D. Ding, T. Chey, E. Stamatakis, W. J. Brown, and A. E. Bauman. 2015. Effect of moderate to vigorous physical activity on all-cause mortality in middle-aged and older Australians. *JAMA Intern Med* 175(6):970–977.

Gholap, N. N., F. A. Achana, M. J. Davies, K. K. Ray, L. Gray, and K. Khunti. 2016, Nov. 12. Long-term mortality following acute myocardial infarction among those with and without diabetes: a systematic review and meta-analysis of studies in the post reperfusion era. *Diabetes Obes Metab*. DOI: 10.1111/dom.12827.

Guigoz, Y., B. Vellas, and P. J. Garry. 1994. Mini nutritional assessment: a practical assessment tool for grading the nutritional state of elderly patients. *Facts Res Gerontol* 2:15–59.

Guligowska, A. R., M. Piglowska, J. Smigielski, and T. Kostka. 2015. Inappropriate pattern of nutrient consumption and coexistent cardiometabolic disorders in elderly people from Poland. *Pol Arch Int Med* 125(7–8):521–531.

Hegewald, J., U. E. Wegewitz, U. Euler et al. 2019. Interventions to support return to work for people with coronary heart disease. *Cochrane Database Syst Rev* 3:CD010748.

Hinman, R. S., M. A. Hunt, M. W. Creaby, T. V. Wrigley, F. J. McManus, and K. L. Bennell. 2010. Hip muscle weakness in individuals with medial knee osteoarthritis. *Arthrit Care Res* 62(8):1190–1193.

Inoue-Choi, M., D. Lazovich, A. E. Prizment, and K. Robien. 2013. Adherence to the World Cancer Research Fund/American Institute for Cancer Research recommendations for cancer prevention is associated with better health-related quality of life among elderly female cancer survivors. *J Clin Oncol* 31(14):1758–1766.

Kiasuwa Mbengi, R. L., A. M. Nicolaie, E. Goetghebeur et al. 2018. Assessing factors associated with long-term work disability after cancer in Belgium: a population-based cohort study using competing risks analysis with a 7-year follow-up. *BMJ Open* 8(2):e014094.

Kostka, J., E. Borowiak, and T. Kostka. 2014a. Nutritional status and quality of life in different populations of older people in Poland. *Eur J Clin Nutr* 68(11):1210–1215.

Kostka, J., J. Czernicki, and T. Kostka. 2014b. Association of muscle strength, power and optimal shortening velocity with functional abilities of women with chronic osteoarthritis participating in a multi-modal exercise program. *J Aging Phys Act* 22(4):564–570.

Kostka, J., M. Niwald, A. Guligowska, T. Kostka, and E. Miller. 2019. Muscle power, contraction velocity and functional performance after stroke. *Brain Behav* 9:e01046:1–7.

Kostka, T., and J. Kostka. 2018. Injuries in sports activities in elderly people. In *Oxford Textbook of Geriatric Medicine*. 3rd ed., eds. Michel, J.-P., B. L. Beattie, F. C. Martin, and J. D. Walston. Oxford, UK: Oxford University Press.

Kostka, T., and K. Bogus. 2007. Independent contribution of overweight/obesity and physical inactivity to lower health-related quality of life in community-dwelling older subjects. *Z Gerontol Geriatr* 40(1):43–51.

Kostka, T., L. M. Arsac, M. C. Patricot, S. E. Berthouze, J.-R. Lacour, and M. Bonnefoy. 2000. Leg extensor power and dehydroepiandrosterone sulfate, insulin-like growth factor-I and testosterone in healthy active elderly people. *Eur J Appl Physiol* 82:83–90.

Larun, L., K. G. Brurberg, J. Odgaard-Jensen, and J. R. Price. 2016. Exercise therapy for chronic fatigue syndrome. *Cochrane Database Syst Rev* (6):CD003200.

Lauzé, M., J. F. Daneault, and C. Duval. 2016. The effects of physical activity in Parkinson's disease: a review. *J Parkinsons Dis* 6(4):685–698.

Leensen, M. C. J., I. F. Groeneveld, I. V. Heide et al. 2017. Return to work of cancer patients after a multidisciplinary intervention including occupational counselling and physical exercise in cancer patients: a prospective study in the Netherlands. *BMJ Open* 7(6):e014746.

Lemura, L. M., S. P. von Duvillard, and S. Mookerjee. 2000. The effects of physical training on functional capacity in adults aged 46 to 90: a meta-analysis. *J Sports Med Phys Fit* 40:1–10.

Li, T., S. Wei, Y. Shi et al. 2016. The dose-response effect of physical activity on cancer mortality: findings from 71 prospective cohort studies. *Br J Sports Med* 50(6):339–345.

Maiorana, A., G. O'Driscoll, C. Cheetham et al. 2000. Combined aerobic and resistance exercise training improves functional capacity and strength in CHF. *J Appl Physiol* 88:1565–1570.

Makary, M. A., D. L. Segev, P. Pronovost et al. 2010. Frailty as a predictor of surgical outcomes in older patients. *J Am Coll Surg* 210(6):901–908.

McCarthy, B., D. Casey, D. Devane, K. Murphy, E. Murphy, and Y. Lacasse. 2015. Pulmonary rehabilitation for chronic obstructive pulmonary disease. *Cochrane Database of Syst Rev* (2). DOI: 10.1002/14651858.CD003793.pub3.

McTiernan, A. 2004. Physical activity after cancer: physiologic outcomes. *Cancer Invest* 22(C):68–81.

Nelson, M. E., W. J. Rejeski, S. N. Blair et al. 2007. American College of Sports Medicine; American Heart Association. Physical activity and public health in older adults: recommendation from the American College of Sports Medicine and the American Heart Association. *Circulation* 116(9):1094–1105.

Norman, K., N. Stobäus, C. Smoliner et al. 2010. Determinants of hand grip strength, knee extension strength and functional status in cancer patients. *Clin Nutr* 29(5):586–591.

Pekmezi, D. W., and W. Demark-Wahnefried. 2011. Updated evidence in support of diet and exercise interventions in cancer survivors. *Acta Oncol* 50(2):167–178.

Pohl, M., G. Rockstroh, S. Rückriem, G. Mrass, and J. Mehrholz. 2003. Immediate effects of speed-dependent treadmill training on gait parameters in early Parkinson's disease. *Arch Phys Med Rehabil* 84(12):1760–1766.

Porter, M. M. 2006. Power training for older adults. *Appl Physiol Nutr Metab* 31(2):87–94.

Rauch, B., C. H. Davos, P. Doherty et al. 2016. The prognostic effect of cardiac rehabilitation in the era of acute revascularisation and statin therapy: a systematic review and meta-analysis of randomized and non-randomized studies – The Cardiac Rehabilitation Outcome Study (CROS). *Eur J Prev Cardiol* 23(18):1914–1939.

Regnaux, J. P., M. M. Lefevre-Colau, L. Trinquart et al. 2015, Oct. 29. High-intensity versus low-intensity physical activity or exercise in people with hip or knee osteoarthritis. *Cochrane Database Syst Rev* (10):CD010203. DOI: 10.1002/14651858.CD010203.pub2.

Robinson, T. N., J. D. Walston, N. E. Brummel et al. 2015. Frailty for surgeons: review of a National Institute on Aging Conference on Frailty for Specialists. *J Am Coll Surg* 221(6):1083–1092.

Rolland, Y., S. Czerwinski, G. Abellan Van Kan et al. 2008. Sarcopenia: its assessment, etiology, pathogenesis, consequences and future perspectives. *J Nutr Health Aging*. 12(7):433–450.

Romero-Ortuno, R., and R. A. Kenny. 2012. The frailty index in Europeans: association with age and mortality. *Age Ageing* 41(5):684–689.

Rørth, R. C. Wong, K. Kragholm, et al. 2016. Return to the workforce after first hospitalization for heart failure. *Circulation* 134(14):999–1009.

Rothstein, J. M., A. Delitto, D. R. Sinacore, and S. J. Rose. 1983. Muscle function in rheumatic disease patients treated with corticosteroids. *Muscle Nerve Suppl* 6(2):128–135.

Rowinski, R., A. Dabrowski, and T. Kostka. 2015. Gardening as the dominant leisure time physical activity of older adults from a post-communist country. The results of the population-based PolSenior Project from Poland. *Arch Gerontol Geriatr* 60:486–491.

Salzwedel, A., R. Reibis, M. D. Heidler, K. Wegscheider, and H. Völler. 2019. Determinants of return to work after multicomponent cardiac rehabilitation. *Arch Phys Med Rehabil* May 2;pii:S0003–9993(19)30288-6. DOI: 10.1016/j.apmr.2019.04.003. [Epub ahead of print].

Schmitz, K. H., K. S. Courneya, C. Matthews et al. 2010. American College of Sports Medicine roundtable on exercise guidelines for cancer survivors. *Med Sci Sports Exerc* 42(7):1409–1426.

Smedegaard, L., A. K. Numé, M. Charlot et al. 2017. Return to work and risk of subsequent detachment from employment after myocardial infarction: insights from Danish nationwide registries. *J Am Heart Assoc* 6(10):e006486. DOI: 10.1161/JAHA.117.006486.

Sołtysik, B. K., Ł. Kroc, M. Pigłowska, A. Guligowska, J. Smigielski, and T. Kostka. 2017. Evaluation of work and life conditions and the quality of life in 60–65 year-old white collar employees, physical workers and unemployed controls. *J Occup Environ Med* 59(5):461–466.

Studenski, S., S. Perera, D. Wallace et al. 2003. Physical performance measures in the clinical setting. *J Am Geriatr Soc* 51(3):314–322.

Swärdh, E., and N. Brodin. 2016. Effects of aerobic and muscle strengthening exercise in adults with rheumatoid arthritis: a narrative review summarising a chapter in Physical activity in the prevention and treatment of disease (FYSS 2016). *Br J Sports Med* 50(6):362–367.

Szybalska, A., K. Broczek, P. Slusarczyk et al. 2018. Utilization of medical rehabilitation services among older Poles – results of the PolSenior study. *Eur Geriatr Med* 9:669–677.

Ullrich, A., H. M. Rath, U. Otto et al. 2017. Outcomes across the return-to-work process in PC survivors attending a rehabilitation measure-results from a prospective study. *Support Care Cancer* 25(10):3007–3015.

Weston, K. S., U. Wisløff, and J. S. Coombes. 2014. High-intensity interval training in patients with lifestyle-induced cardiometabolic disease: a systematic review and meta-analysis. *Br J Sports Med* 48(16):227–1234.

Williams, P. T. 2013. Inadequate exercise as a risk factor for sepsis mortality. *PLoS One* 8(12):e79344.

Williams, P. T., and P. D. Thompson. 2014. Increased cardiovascular disease mortality associated with excessive exercise in heart attack survivors. *Mayo Clin Proc* 89(9):1187–1194.

Wilson, P. W., R. B. D'Agostino, L. Sullivan, H. Parise, and W. B. Kannel. 2002. Overweight and obesity as determinants of cardiovascular risk: the Framingham experience. *Arch Intern Med* 162(16):1867–1872.

Wolin, K. Y., A. L. Schwartz, C. E. Matthews, K. S. Courneya, and K. H. Schmitz. 2012. Implementing the exercise guidelines for cancer survivors. *J Support Oncol* 10(5):171–177.

Wolvers, M. D. J., M. C. J. Leensen, I. F. Groeneveld, M. H. W. Frings-Dresen, and A. G. E. M. De Boer. 2018. Predictors for earlier return to work of cancer patients. *J Cancer Surviv* 12(2):169–177.

Wysokiński, A., T. Sobów, I. Kłoszewska, and T. Kostka. 2015. Mechanisms of the anorexia of aging – a review. *Age (Dordr)* 37(4):9821.

Zheng, G., W. Zhou, R. Xia, J. Tao, and L. Chen. 2016. Aerobic Exercises for cognition rehabilitation following stroke: a systematic review. *J Stroke Cerebrovasc Dis* 25(11):2780–2789.

7 Physical and Psychosocial Work Demand Changes with Age

Joanna Bugajska
Central Institute for Labour Protection –
National Research Institute

Teresa Makowiec-Dąbrowska
Nofer Institute of Occupational Medicine

CONTENTS

DEFINITION OF AGING WORKER

Ageing is an irreversible process for every human being; however, the course of this process involves significant intra- and interpersonal changes. Intrapersonal changes are the processes in which age-related decline does not equally affect all the functions responsible for work ability; hence, the extent of the aforementioned decline does not encompass all the areas of professional life. Therefore, despite advancing age, the levels of individual ability to do certain types of work may still be high. The range of interpersonal differences increases with age, since the pace of ageing process is person-specific.

This issue is still more complex due to the fact that the ability to perform work is modified by other features that are acquired with age, such as high experience or higher emotional resistivity. In older workers, the above features can often compensate the effects of age-related decline in physical capacity and fitness levels, as well as some psychophysical functions.

Determining the age when the ageing process begins and work ability declines is a very complex process. Setting the cut-off point when a worker becomes "an older worker", based on his/her chronological age, is inadequate as regards occupational activity performance. In numerous cases, the period of efficient functioning is proportional to life span, with no specific signs that are usually attributed to old age. Therefore, determining the so-called functional age, based on the measurement of all the parameters reflecting physical and mental activity, not on chronological age, would be a better approach. Such an assessment considers individual differences in the ageing process and is in conformity with the recommendations for aged individuals, issued by the International Labor Organization (ILO No 162 1980). This document defines older individuals as persons who, due to their age, can encounter problems regarding both employment and work performance at any age. However, as regards the definition of an "older worker", it is essential to determine the moment when the age-related decline in certain indicators of psychophysical capacity may result in impaired ability to do specific work, posing threat to the worker's and other people's health (Koradecka et al. 2006).

Attempts have been made to establish this moment, based on the assessment of various functions of the human body, such as the following:

- According to Nygard, after 50 years of age, a decline is noted in musculo-skeletal system efficiency (maximal isometric strength of dorsal muscles decreases by 16%–22% in men and by 9%–10% in women), which may result in excessive load applied to workers at this age, who perform hard physical work (Nygård et al. 1991).
- According to Ilmarinen, work ability evidently declines after 50 years of age irrespective of individual functional differences. This limitation is due to a marked decrease in physical capacity and impaired dorsal muscle strength in women and men aged 50 years and above (Ilmarinen 1988).

Discussions about the effect of ageing on work ability in workers are mainly focused on identifying the factors present in working environment, which can be burdensome for older workers. An interesting aspect concerning the relationship between working conditions and ageing has been reported by Freude and co-workers (2009). The authors carried out a study on the differences between the chronological age and biological age, measuring such parameters as arterial blood pressure (ABP), resting and exercise heart rate (HR), muscle strength, vision, hearing, reaction time, memory, concentration, physical and functional complaints, and social functioning. Three hundred and seventy-one persons aged 20–64 years participated in the study. The participants belonged to five groups, namely teachers (women and men), clerks (women), nursing school teachers (women) and executives (men). The biggest differences were found in the parameters obtained from executives (the difference between their chronological and biological age was 9 years in favour of the latter), next in female teachers (the difference was 5 years) and in nursing school teachers (the difference was about 2 years). The smallest difference (about 2 years) was noted in female clerks (office workers). Based on the multifactor regression analysis, it was found that the predictors of bigger differences between chronological and biological age include mental resources (WAI 7), as well as earnings and the percentage of body fat mass (Freude et al. 2009). Regretfully, the authors did not assess the effect of other factors pertaining to working environment or job requirements (especially physical requirements) and did not consider the population of physical workers in their study.

Although it is difficult to univocally state at what age a worker becomes "an older worker", setting this criterion is crucial from the point of view of statistics, reports, or epidemiological studies tracing work-related activity and health status in this population of workers. It would also help to identify the target groups for the implementation of preventive programs for older workers. The WHO expert panel, in their report on the relationship between aging and working capacity, concluded that 45+ is the age when a worker becomes "an older worker" (WHO 1993). Conversely, according to the reports on regular surveys related to working conditions in EU member states, which have been regularly carried out since 1990 by the European Agency for Working and Life Conditions, 50+ (50 years and above) is the age when a worker is referred to as "an older worker".

DISPARITIES BETWEEN WORKING CONDITIONS DUE TO WORKERS' AGE ACCORDING TO THE SIXTH EUROPEAN WORKING CONDITIONS SURVEY

Men's work ability changes with the ageing process. This is due to the decline in physical capacity and fitness levels, and some components of psychophysical fitness (e.g. perception, reaction time and sensory capacity). Simultaneously, the prevalence of chronic conditions of the circulatory, respiratory and musculoskeletal systems, as well as endocrinal and digestive disorders, is higher in older population. The job requirements, in turn, unless the worker changes his/her job, usually remain the same regardless the worker's age. All of the above-mentioned factors result in work-load increase, which is proportional to the worker's age.

The data presented next reveal the disparity between working conditions considering workers' age. These parameters correspond to the incidence of selected psychophysical and organizational factors in the working environment, according to the subjective assessment of working conditions using the questionnaire of the European Foundation for the Improvement of Living and Working Conditions in Dublin. Presently, our knowledge of this issue is based on the data obtained from a representative sample of workers from 28 EU member states, according to the Sixth European Working Conditions Survey 2015 (EWCS 2015). These data indicate that there are no big differences between the job requirements towards younger (below 35 years of age) and older (50+) workers (Table 7.1). The data suggest that, for some workers, physically demanding work is a major characteristic of their job:

- The activity involving repetitive hand or arm movements within a quarter of the time or more is the most dominant job-related risk factor, affecting more than 60% of workers (63% of workers aged below 35 years 35, and 58% of workers aged 50 years and above).
- Tiring or painful positions assumed within a quarter of the time or more are reported by 43% at all of workers (42% of workers aged below 35 years, and 43% of workers aged 50 years and above).
- A third or more of workers report that their work requires them to carry and move heavy loads a quarter of the time or more (35% of workers aged below 35 years, and 29% of workers aged 50 years and above).

TABLE 7.1
Working Conditions according to Age in EU 28 and in Poland (Percentage of Exposed Workers) - Sixth EWCS 2015)

Exposed to:	Age (Years)	EU 28 2015	Poland 2015
Carrying or moving heavy loads	<35	35	33
	35–49	32	37
	50+	29	35
	All	32	35
Lifting and moving people	< 35	10	7
	35–49	10	9
	50+	9	10
	All	10	9
Work involves repetitive hand or arm movements	<35	63	59
	35–49	62	62
	50+	58	57
	All	61	59
Work involves tiring or painful positions	<35	42	45
	35–49	44	47
	50+	43	47
	All	43	46
More than 40 hours per week in main paid job	<35	20	31
	35–49	26	32
	50+	23	26
	All	23	30
High temperatures	<35	24	28
	35–49	24	31
	50+	23	27
	All	23	29
Low temperatures	<35	22	24
	35–49	21	25
	50+	22	29
	All	21	25
Vibrations from tools or machinery	<35	21	31
	35–49	21	26
	50+	19	23
	All	20	27
Working to tight deadlines	<35	66	60
	35–49	67	63
	50+	59	47
	All	64	57
Shift work	<35	26	31
	35–49	22	32
	50+	16	26
	All	21	30

Based on the comparison of the results obtained in workers aged 50 years and above in Poland with the results obtained from the total 50+ worker population in the EU, we can generally conclude that Polish workers are most often required to carry or move heavy loads. They more often assume tiring or painful positions and are exposed to high and low temperatures, and vibrations from tools or machinery. They more often work shifts. Conversely, they are more seldom required to work to tight deadlines (Table 7.1).

As regards the organizational risk factors, "working to tight deadlines" is most frequently mentioned by workers. It is reported by 64% of workers at all (66% of workers aged below 35, and 59% of workers aged 50 years and above).

Sixty-one per cent of workers 50+ (50 years and above) believe that work has no effect on their health, whereas 14% report mainly favourable effects of work on their health. A slightly higher percentage of workers aged below 35 years (67%) believe that their work does not affect their health, whereas only 11% report a favourable effect of work on their health.

No significant differences have been found in absence from work for health reasons within the last 12 months between the two age groups. Fifty-five per cent of the workers aged 50 years and over report no absence from work as compared to 54% of workers aged below 35 years. At the same time, the percentage of workers aged 50 years and above, who have reported going to work despite poor health, is lower (39%) as compared to the value obtained from the group of workers aged below 35 years (44%).

Significant differences are noted in answers to the question "Do you think you will be able to do your current or a similar job until you are 60 years old?". In as many as 78% workers aged 50 years and above, the answer to this question was "yes", whereas in the group of workers aged below 35 years, 62% believe they will be able to do a similar job when they are 60 years old.

DIFFERENCES IN WORKING CONDITIONS ACCORDING TO AGE IN RESEARCH

The issue of unchanging working requirements, irrespective of the worker's ageing, is recognized by many researchers and has been the goal of multiple studies on age-associated changes in working conditions and workload in persons of different age, working under the same conditions. Louhevaara (1999), based on the observation of two occupational groups (construction workers and transport inspectors), has found that the requirements relating to work performance remain unchanged despite the worker's age, and thus, the workload involving physical and static effort of older workers may be greater than that of their younger counterparts. Construction workers are periodically exposed to very high levels of physical effort, which can result in increased HR values, reaching the level close to the maximal HR, considering the worker's age (Louhevaara, 1999). This also concerns persons doing jobs which are believed to be extremely burdensome for the worker, such as fireman's job. Bugajska in her study on workload in firemen has found that during typical occupational activities such as climbing up the ladder or transporting the victims, the mean HR values reach 99% of the maximal value, considering the worker's age according to the established PN-EN 9669 norm (Bugajska et al. 2007). Based on the results obtained by Sluiter and Frings-Dresen (2007) in Dutch firemen, significant between-subject and between-group differences (obtained from various age groups) were found in responses from the circulatory system during a fire alarm and rescue or fire extinguishing operations. In firemen aged 50 years and above, the HR values may reach the levels from 4% to 81% of heart rate reserve (HRR), which is a difference between resting HR values and d maximal HR, while during rescue and fire extinguishing operations, HR values reach the levels from 10% to 56% (Sluiter and Frings-Dresen 2007). Such high loads applied to the cardiovascular system may result in sudden cardiac events, especially in older firemen (Peate et al. 2002).

Accordingly, the study carried out by Makowiec-Dąbrowska among postmen and postwomen has shown that their workload during field work gradually increases with age (Makowiec-Dąbrowska 2012). In persons aged above 40 years, the HR value exceeded 60% HRmax, indicating that the work is excessively burdensome for older workers (Figure 7.1).

Furthermore, while in persons aged below 40 years, HR values decreased during subsequent working hours with the decrease of mailbag weight, in older people an increase in HR values indicating accumulating fatigue was observed after the second working hour (Figure 7.2).

The problem with the feeling of fatigue in older workers and the associated prerequisite for changes in work organization is also reported in other papers. The results of the studies performed among public sector workers indicate a growing need for recovery after work. It is especially evident in full-time workers doing a monotonous work under time pressure. The need for breaks is also more evident in women and persons with symptoms from the musculoskeletal system. This need is

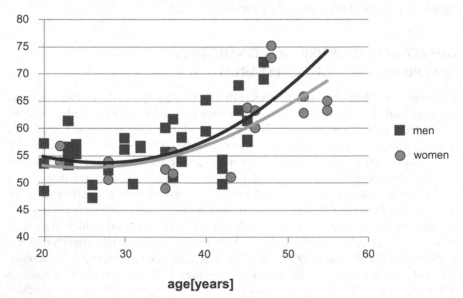

FIGURE 7.1 HR (as % HRmax) during carrying mailbags and the postmen's and postwomen's age.

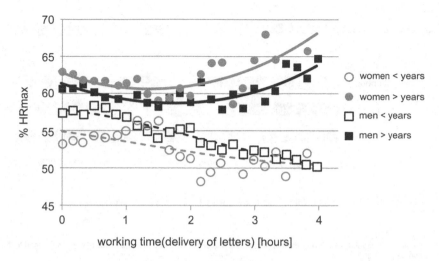

FIGURE 7.2   HR (as % HRmax) during subsequent hours of mail service, considering the postmen's and postwomen's age.

not so pressing, however, when social support is provided by the management (Kiss et al. 2008).

Considering the age-related changes in the human body, which are responsible for decreasing physical and adaptive capacity, the workload imposed on older workers should be reduced. The decreasing share of older people among those who work hard or work in demanding work enviroment is the evidence of the implementation of this postulate. To check whether and to what extent this problem is reflected in Polish conditions in a group of employees who were subject to an analysis of the impact of individual, professional and non-professional factors on the subjectively perceived ability to work (as detailed in Chapter 4).

The results of this analysis, separately for men and women, are presented in the following figures. In the group of women, a positive situation (fewer people aged 50 years and above exposed to a given aggravating factor compared to non-exposed) was found according to moderate and strenuous work (Figure 7.3), exposure to chemical factors (Figure 7.4) and working in 12-hour shifts or 24-hour duty (Figure 7.5). On the other hand, the completely opposite situation (more people aged 50 years and above exposed to a given aggravating factor compared to the non-exposed) concerned exposure to incorrect lighting (Figure 7.4), highly repetitive task, uneven work pace, a large amount of labour and day work (Figure 7.5).

In the group of male workers, no differences related to energy expenditure were found in workers aged 50 years and above, (Figure 7.6). However, there were more workers aged 50 years and above among those working in exposure to chemical factors, very high temperatures and mineral dust, as compared with the unexposed ones (Figure 7.7). Conversely, fewer workers aged 50 years and above exposed to burdensome factors, as compared to unexposed ones, were noted in case of exposure to whole-body vibration, a large amount of labour, and working according to 12-hour shift or 24-hour duty schedules (Figures 7.7 and 7.8).

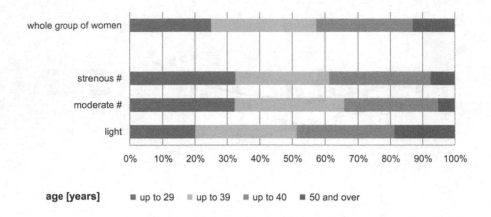

statistically significantly lower percentage of persons over 50 years of age in the exposed group compared to the non-exposed group

FIGURE 7.3 Age distribution in the entire sample of females and subgroups selected according to energy expenditure at work.

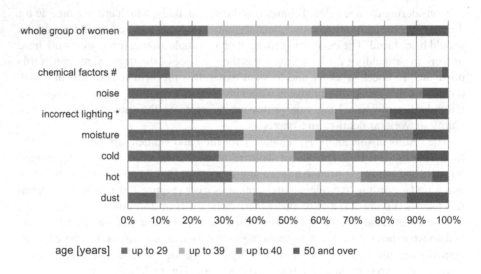

* statistically significantly higher percentage of persons over 50 years of age in the exposed group compared to the non-exposed group
# statistically significantly lower percentage of persons over 50 years of age in the exposed group compared to the non-exposed group

FIGURE 7.4 Age distribution in the entire sample of females and subgroups, according to individual burdensome factors present in their working environment.

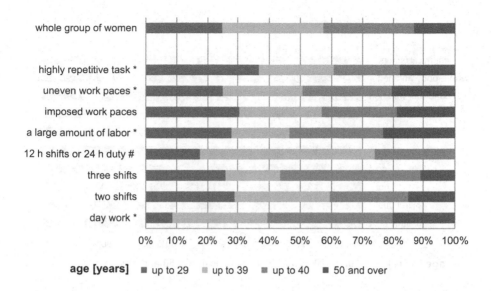

* statistically significantly higher percentage of persons over 50 years of age in the exposed group compared to the non-exposed group
# statistically significantly lower percentage of persons over 50 years of age in the exposed group compared to the non-exposed group

FIGURE 7.5   Age distribution in the entire sample of females and subgroups selected according to individual inconveniences resulting from work organization.

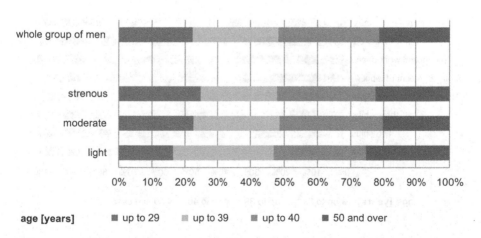

FIGURE 7.6   Age distribution in the entire sample of males and subgroups of subjects selected according to energy expenditure at work.

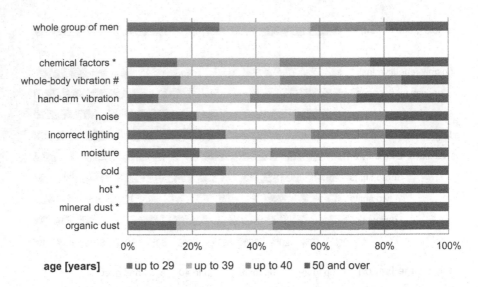

FIGURE 7.7 Age distribution in the entire sample of males and subgroups exposed to individual burdensome factors in their working environment.

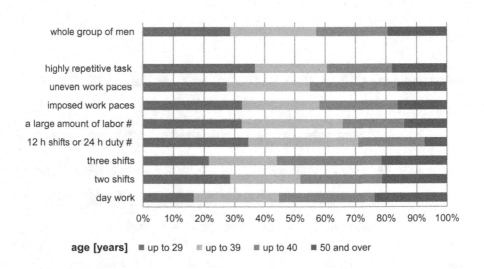

FIGURE 7.8 Age distribution in the entire sample of males and subgroups selected according to exposure to individual burdensome factors resulting from work organization.

Such changes in the number of older workers among the working population exposed to potentially burdensome factors entailed some consequences. In each case of a lower percentage of workers aged 50 years and above in the working population exposed to individual burdensome factors, a decrease was noted in the values of negative correlation coefficient and Work Ability Index (WAI), or the correlation was positive. The correlation coefficient for age and WAI in hard-working female physical workers were $r = 0.032$ and $p = 0.707$, whereas the corresponding values obtained in female workers with different energy expenditures were $r = -0.200$ and $p = 0.000$, respectively. It doesn't mean that hard physical work prevents the age-related decrease in subjective work ability rating. On the contrary, it indicates that older workers are not allowed to perform a hard physical work. A larger number of persons aged 50 years and above among the workers exposed to burdensome factors resulted in age-related decline in work ability, expressed by WAI parameters. In the male population exposed to mineral dust, the correlation between age and WAI was moderately negative (($r = -0.546$, $p = 0.004$), whereas in the group of unexposed workers, it was also negative, yet low ($r = -0.266$, $p = 0.000$).

CONCLUSIONS

Summing up the data obtained from EWCS and the research results, we may conclude that changes in working conditions, involving workload reduction in older workers, are not commonplace. Despite the age-related decrease in work ability, the demographic and economic determinants indicate the need for prolonged periods of occupational activity. Encouraging older people to continue work is justified only in cases when parallel solutions are implemented. They should be focused on maintaining work ability in workers within the entire period of their occupational activity and on modification of tasks, workstations, working time and rhythm, etc. to adjust them to age-related changes in workers' potential. A systemic implementation of such solutions is called age management. Implementation of the tasks identified in age management systems should be mainly based on the knowledge of individual and work-related factors fostering prolonged workers' work ability within the entire range of their occupational activity. These goals can be achieved through the development of programs and policies in four domains, namely working environment, work organization and work/life balance, health promotion, disease prevention and social support (Silverstein 2008).

REFERENCES

6-th European Working Conditions Survey (EWCS). 2015. https://www.eurofound.europa.eu/data/european-working-conditions-survey (accessed September 18, 2019).

Bugajska, J., K. Zużewicz, M. Szmaus-Dypko, and M. Konarska. 2007. Cardiovascular stress, energy expenditure, and subjective ratings of fire fighters during typical fire suppression and rescue tasks. *Int J Occup Saf Ergon* 13(3):323–332. DOI: 10.1080/10803548.2007.11076730.

Freude, G., O. Jakob, P. Martus, U. Rose, and R. Seibt. 2010. Predictors of the discrepancy between calendar and biological age. *Occup Med (Lond)* 60(1):21–28. DOI: 10.1093/occmed/kqp113.

Ilmarinen, J. 1988. Physiological criteria for retirement age. *Scand J Work Environ Health* 14 (Suppl 1):88–89.

ILO. 1980. Older workers recommendation, (No. 162). ILOLEX Database of International Labour Standards; ILO.

Kiss, P., M. De Meester, and L. Braeckman. 2008. Differences between younger and older workers in the need for recovery after work. *Int Arch Occup Environ Health* 81(3):311–320. DOI: 10.1007/s00420-007-0215-y.

Koradecka, D., J. Bugajska, and Z. Pawłowska. 2007. Merytoryczne przesłanki do projektu ustawy o emeryturach pomostowych. *Bezpieczeństwo Pracy - Nauka i Praktyka* 10:2–6.

Louhevaara, V. 1999. Is the physical work load equal for ageing and young blue-collar workers. *Int J Ind Ergonom.* 24(5):559–564.

Makowiec-Dąbrowska, T. 2012. Tolerancja wysiłku fizycznego w pracy zawodowej a wiek pracowników. Referat: XII Krajowy Zjazd Naukowy Polskiego Towarzystwa Medycyny Pracy. Poznań, 12–15 wrzesień 2012 r.

Nygård, C.-H., I. Eskelinen, S. Suvanto, K. Tuomi, and J. Ilmarinen. 1991. Associations between functional capacity and work ability among elderly municipal employees. *Scand J Work Environ Health* 17(1):122–127.

Peate, W. F., L. Lundergan, and J. J. Johnson. 2002. Fitness self-perception and VO_2max in firefighters. *J Occup Environ Med* 44(6):546–550. DOI: 10.1097/00043764-200206000-00017.

Silverstein, M. 2008. Meeting the challenges of an aging workforce. *Am J Ind Med* 51(4):269–280. DOI: 10.1002/ajim.20569.

Sluiter, J. K., and M. H. W. Frings-Dresen. 2007. What do we know about ageing at work? Evidence-based fitness for duty and health in fire fighters. *Ergonomics* 50(11):1897–1913. DOI: 10.1080/00140130701676005.

WHO Study Group on Aging and Working Capacity & World Health Organization. 1993. Aging and working capacity: report of a WHO study group [meeting held in Helsinki from 11 to 13 December 1991]. World Health Organization. https://apps.who.int/iris/handle/10665/36979.

8 Work Ability in Older Women Working in Retail Sector – Results of Research

Elżbieta Łastowiecka-Moras
Central Institute for Labour Protection –
National Research Institute

CONTENTS

INTRODUCTION

In the era of mechanization and automation, the issue of physical work is often neglected due to the common perception that its physical impacts are less and less frequent. Paradoxically, in recent years, this problem has become more serious, and, in a number of cases, it has been identified as the consequence of technological progress (Koszada-Włodarczyk et al. 2001). The introduction of new technologies, which are mainly focused on increasing work productivity and the automation of selected sections in production lines, has led to a faster pace of work, increased quantities of goods and higher frequency of operations performed by workers, all of which have an impact on workload levels.

A lot of physically demanding work is found in the retail trade sector, which is an important area of economic life. Trade business organizations generate 17% of gross added value, whereas sectors indirectly associated with commerce produce additional 12%. As a result, trade contributes to the generation of as much as 29% of added value in the economy (Kierunki 2014. Made in Poland 2014).

In 2018, there were 106,000 active shops and commercial sites in the Polish market (GUS 2019b).

The retail trade sector is vast, differentiated and constantly growing, offering numerous jobs. Thus, persons of any age with any education level may find jobs in trade. The commercial sector in Poland offers more than 2 million jobs, which makes 15% of the total labour force (Momot 2016).

It is of note, however, that as many as 40% of employees in the retail trade sector are above 50 years of age and that a vast majority (70%) of these employees are women (GUS 2019a). The only available workplaces that are suitable for these female employees often require physical effort exceeding their physical abilities.

Female physiology prevents some women from being equally productive at work as men (Kozłowski and Nazar 1995). In the case of older women, in addition to physiology and gender-based differences, the age-related functional changes significantly compromise their physical ability. In older women, these age- and gender-based conditions lead to suffering from excessive workloads in jobs requiring manual labour and contribute to various lesions, especially in the musculoskeletal, cardiovascular and nervous systems (Augustyniuk et al. 2016).

As estimated by trade workers, particularly the female ones, jobs in retail trade sector are physically demanding. Obviously, the workload in this sector depends on multiple factors including types of stores (grocery stores vs. non-food stores) or the commercial space of a given retail unit.

In smaller shops employing few workers, there are usually no job specialties, and each worker performs duties of a shop assistant and a teller at the same time.

Apart from the usual activities related to customer service and teller duties, retail sales workers have to unload goods and pack the items on shelves, clean store facilities and do other works as required. Since they have to combine customer service with ancillary works, their workload is uneven, which is beneficial due to overall lower static load imposed on their bodies at the workstation. The range of duties of large format store workers is usually narrower. The tellers mostly stick to handling cash transaction. This work is usually performed in a seated position, sometimes forced, often during the better part of their working shift, which can be detrimental from the point of view of ergonomics. Unfortunately, due to staff shortages, the super- and hypermarket staff members increasingly often have to perform extra duties (e.g. cleaning), which results in over-normative workload. In consequence, the workers are not prepared for such tasks and are at risk of numerous symptoms, mainly from the musculoskeletal system.

Due to the increase in the purchasing power, increasing supply for goods and cumulative supply of assortment in large retail trade units, the total mass of the goods lifted by, e.g., a female teller reaches even several tons daily (Bortkiewicz et al. 2011). Physical work, especially when it requires effort beyond the worker's capacity, may lead to a decrease in work ability. Research results indicate that such factors as very hard work, repeatable activities, a forced, uncomfortable body position and working overtime are the significant independent factors responsible for a premature complete loss of work ability.

The aim of the study was to assess work ability among women over 55, involved in physical work in the retail sales sector including commercial sites of various selling areas.

MATERIALS AND METHODS

STUDY GROUP

Three hundred women aged 55 years and above, employed in the retail trade sector at a workplace involving manual labour with at least five years of work experience in the above, participated in the study.

METHODS

The study used a survey involving a direct personal interview using Paper and Pen Personal Interview (PAPI). The questionnaire included closed questions, multiple choice questions and open questions allowing unrestricted answers. The survey was carried out by trained surveyors in a selected representative sample. Based on the criterion of physical footprint, the retail units were categorized into large format stores occupying a large physical space, exceeding 300 square meters and small format stores occupying smaller spaces up to 300 square meters (Andrzejczak et al. 2010; Pasternak and Musiał 2008).

The large format stores included supermarkets, hypermarkets and discount stores. The small format retail segment, in turn, comprised big, medium size and small grocery stores, chemist shops, pharmacies and specialist grocery stores. Selection of the sample was based on quota and judgmental sampling where the quotas were determined based on the employment in large format stores or small format shops. The samples were equally divided. Using this approach, the equipotent sample was obtained, divided according to the site of employment.

According to the accurately determined selection pathway, a direct survey was conducted by a trained surveyor. The questionnaire developed for the study contained questions relating to personal data and occupational activities, including work ability in the studied sample of women. The questions related to working conditions were taken from the questionnaire applied during the recent European study on working conditions (EWCS 2015). Since 1990, the European Foundation for the Improvement of Living and Working Conditions based in Dublin carries out cyclic research on working conditions, on the basis of the work-related risk subjective assessment carried out in EU countries. Owing to the uniform methodological approach, it is possible to compare the results of research in each member state participating in the research.

This study used questions pertaining to the following issues: employment status, working time duration and organization, work organization, and physical and psychosocial risk factors.

The *Work Ability Index (WAI)* was used for work ability measurements. Work ability is a complex feature. Its level results from interactions between job requirements, the degree of physical and mental effort in accordance with the worker's functional capabilities, health status and self-assessment of functioning in a given organizational and social setting. Ability to work may be assessed objectively, i.e., by analysing the functional capacity of a given worker plus his/her skills and abilities, or subjectively, i.e. based on the evaluated person's opinion. Such an approach to work ability rating was proposed at the Finnish Institute of Occupational Health in Helsinki, where the so-called WAI was developed (Tuomi et al. 1998).

STATISTICAL ANALYSIS

The feedback including 300 questionnaires was verified for correctness of answers, coded and next uploaded to the computer database. Statistical analysis of the results was conducted using SPSS Software version 15.

The data obtained during the research were next subjected to the multilevel quantitative and qualitative analysis. This chapter presents the results corresponding to general trends, the mean values obtained from the entire sample and the selected independent variables. For the purpose of the research, the following independent variables were selected: workplace (large format stores and small format stores), age, main assortment sold at the retail unit (grocery stores and other stores) and working hours (full-time, part-time work, etc.). As regards the aforementioned variables, the values reflecting the highest and the lowest levels of a given independent variable were compared as well as the decreasing and increasing trends in the internal structure of the responses. The subgroups with the highest and the lowest percentage of answers to a given question were also described. Selection of such independent variables is extremely important from the point of view of their correlations with the total values obtained from the entire studied population. Nonparametric Mann–Whitney U-test was used to compare the two independent samples, whereas Kruskal–Wallis test was applied for multiple sample comparison.

RESULTS

The results of the study are presented for the entire group of respondents and for the size of trade sites where the surveyed women had been employed.

GENERAL DESCRIPTION OF THE STUDIED SAMPLE

The sample included 300 women aged 55–72 years (the mean age of the participants was 57.8 ± 2.51 years). Out of the entire studied population, the majority of female respondents were women with secondary (43.8%) and basic vocational (41.5%) education. Table 8.1 presents the anthropometric data of the surveyed women.

PROFESSIONAL ACTIVITY OF THE PARTICIPANTS

Small format stores employed 174 women (58%) and large format stores – 126 women (42%).

TABLE 8.1
Anthropometric Data of the Studied Group of Women (N = 300)

Parameter	N	Mean Value	SD	Min–Max
Body weight (kg)	300	71.3	11.45	48–102
Body height (cm)	126	166.9	5.83	149–185
BMI	126	25.6	3.85	17.96–35.71

The vast majority (70%) of the respondents was employed at grocery shops.

The most frequent positions held by the surveyed women included shop assistants – 210 women (70%), tellers – 58 (19.3%) and unskilled physical workers – 26 (8.7%).

The average value corresponding to total work experience in the studied sample was 32.52 years SD ± 6.71 and ranged from 12 to 52 years. Table 8.2 presents information on work experience in the studied sample according to store footprint.

Table 8.3 presents the average weekly working time of women depending on the footprint of the retail unit.

A statistically significant higher number of weekly working hours were noted in the female participants working in large format stores. They were found to work at night and in evenings, more often than the women working in small format stores.

Almost half of the participants worked in a two-shift system (working two shifts with a break at night and at the end of the week), whereas one-third and 12% of the sample worked one shift and three shifts (three-shift work with a break at the end of the week), respectively.

The majority (66.7%) of the surveyed women performed diverse tasks at work, and most of their tasks required physical effort. The remaining 33.3% of the respondents performed a typically physical work.

TABLE 8.2
Work Experience (in Years) of the Studied Sample (N = 300)

	Large Format Stores N = 126			Small Format Stores N = 174			
	Mean Value	SD	Min–Max	Mean Value	SD	Min–Max	p
Total	31.16	6.68	12–45	33.51	6.58	19–52	>0.05
In retail trade	17.05	7.26	5–40	17.8	8.9	5–40	>0.05
In current job	7.39	4.73	1–34	7.43	4.37	1–30	>0.05

TABLE 8.3
Working Time in the Studied Group of Females (N = 300)

Parameter	Large Format Stores	Small Format Stores	p
Number of weekly working hours	43.6	41.3	0.018
Number of monthly working Saturdays	2.16	2.36	>0.05
Number of monthly working Sundays	1.79	1.94	>0.05
Number of monthly working nights (at least 2 hours between 10.00 p.m. and 5.00 a.m.)	8.83	6.16	0.039
Number of monthly working evenings (at least 2 hours between 6.00 p.m. and 8.00 p.m.)	8.88	7.53	0.027

Exposure to Dangerous, Harmful and Burdensome Factors at the Workplace

Table 8.4 presents the mean time of exposure to dangerous, harmful and burdensome factors at the workplace in the entire sample of women according to store footprint.

Statistically significant higher values corresponding to the time of lifting and/or handling heavy objects up to and above 10 kg using assistive devices and the time of repetitive work were obtained in women working in large format stores as compared to their counterparts working in small format stores. This data indicates generally higher workload in large format retail units.

Work Ability in the Surveyed Group of Women

Work Ability Index – Total Score

The mean WAI value obtained in the group of studied women ranged from 15 to 49 with its mean value of 38.52 (SD ± 6.59), ranking it in the category of good ability to work. Table 8.5 presents the comparison of global WAI values according to age, store footprint, assortment and job.

The presented data indicates that younger women obtained higher scores for their work ability rating as compared to those aged above 50 years; accordingly, higher scores were obtained by the women working in small format shops as compared to those working in large format shops from women working in stores selling non-food products as compared to those working in grocery stores and from the female sales assistants as compared to the values obtained from tellers and unskilled physical workers.

Next, the answers to particular questions of the WAI questionnaire were analysed according to store size.

TABLE 8.4
The Mean Time of Exposure to Dangerous, Harmful and Burdensome Factors (in hours) at the Workplace During the Working Shift, in General and According to Store Footprint

	Factor	Large Format Stores	Small Format Stores	p
1.	Inhaled chemicals, gases or fumes	1.38	1.37	>0.05
2.	Inhaled dusts	1.35	1.26	>0.05
3.	Skin contact with chemicals	2.41	1.57	>0.05
4.	Cigarette smoke from the smoking workers	1.56	1.33	>0.05
5.	Contact with materials that can be infected, e.g. waste (litter) or saline	1.79	1.29	>0.05
6.	Contact with fruit and vegetable preservatives	2.63	1.49	>0.05
7.	Contact with toxic insecticides and rodenticides	1.73	1.40	>0.05
8.	Noise so loud that you would have to raise your voice to talk to people	2.19	1.28	>0.05
9.	Noise which is not very loud, yet still burdensome	3.26	2.07	>0.05

(Continued)

TABLE 8.4 (*Continued*)

The Mean Time of Exposure to Dangerous, Harmful and Burdensome Factors (in hours) at the Workplace During the Working Shift, in General and According to Store Footprint

	Factor	Large Format Stores	Small Format Stores	p
10.	Vibration of machines and other devices, affecting arms, legs or the trunk	1.53	1.29	>0.05
11.	High temperatures causing sweating even when there is no work to do (exposure to sunlight from outside and heat from freezers)	2.44	1.58	>0.05
12.	Low temperatures in store facilities	2.20	2.03	>0.05
13.	Low temperatures outside the facilities	1.69	1.66	>0.05
14.	High humidity levels in facilities	1.93	1.58	>0.05
15.	Inadequate lighting (insufficient illuminance, glare)	3.88	2.91	>0.05
16.	Ultraviolet radiation	1.07	1.35	>0.05
17.	Electromagnetic fields	1.20	1.21	>0.05
18.	Exposure to electric shock	1.62	1.15	>0.05
19.	Explosion and/or fire	1.67	1.38	>0.05
20.	Injuries caused by mobile parts of machines	1.93	1.27	>0.05
21.	Injuries due to being struck by falling elements and goods in piles	2.82	1.62	>0.05
22.	Exposure to cuts and abrasive lesions caused by sharp or harsh edges or broken glass	3.06	2.13	>0.05
23.	Falling while walking on slippery or uneven surfaces	2.47	1.64	>0.05
24.	Falling from ladders, landing, etc.	2.13	1.45	>0.05
25.	Substantial physical effort causing fatigue	3.77	2.53	>0.05
26.	Contact with hot surfaces	1.45	1.47	>0.05
27.	Contact with cold surfaces	1.56	1.41	>0.05
28.	Working in a seated position within a long period of time, resulting in fatigue and/or pain	3.37	2.06	>0.05
29.	Working in a standing position within a long period of time, causing fatigue and/or pain	4.13	3.11	>0.05
30.	Bending and trunk rotation	3.10	2.28	>0.05
31.	Squatting and kneeling causing fatigue and/or pain	2.69	1.95	>0.05
32.	Working with arms raised overhead	2.69	1.51	>0.05
33.	Lifting and/or carrying objects up to 10 kg	3.43	2.58	<0.05
34.	Lifting and/or carrying objects above 10 kg.	3.10	1.78	<0.05
35.	Handling heavy objects	3.23	2.16	>0.05
36.	Pushing and/or pulling heavy objects using assistive devices (trolleys, wheelbarrow)	3.28	1.54	<0.05
37.	Performing repetitive movements of arms and/or shoulders (monotype)	4.71	2.76	<0.05
38.	Contacts with persons other than the personnel members, e.g. customers	6.71	5.62	>0.05

TABLE 8.5
WAI Global Value by Age, Store Footprint, Assortment and Job

Parameter	Work Ability				
	Mean Value	Min	Max	SD	p
Total	38.52	15	49	6.59	
Large format stores	36.83	15	48	7.04	<0.05
Small format stores	39.75	17	49	5.98	
55–59 years of age	39.05	15	49	6.50	<0.05
60+ years of age	36.96	15	49	6.66	
Alimentary products	37.90	15	48	6.77	<0.05
Non-food products	39.98	25	49	5.95	
Sales assistant	39.26	15	49	6.04	<0.05
Teller	37.24	15	48	7.38	
Physical worker	35.81	17	47	7.77	

CURRENT WORK ABILITY COMPARED WITH LIFETIME BEST

Figure 8.1 presents the mean score corresponding to the first question of WAI questionnaire.

No statistically significant differences related to store size were noted in the score corresponding to the first question of WAI questionnaire.

WORK ABILITY IN RELATION TO THE DEMANDS OF THE JOB

Figure 8.2 presents answers to question no. 2, corresponding to physical demands of the work. The values are compared according to store size.

Figure 8.3 presents the answer to question no. 2 of WAI, referring to mental demands of the job. The values are compared according to store size.

Higher scores were obtained from the required mental effort capacity assessment as compared to the required physical effort capacity assessment.

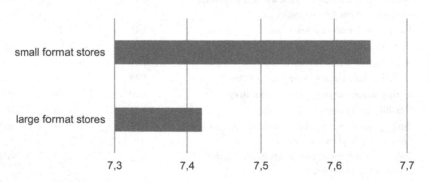

FIGURE 8.1 Current work ability as compared to lifetime best (top form) according to store size.

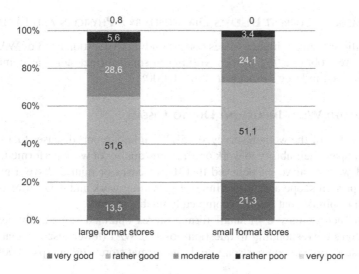

FIGURE 8.2   How do you rate your current work ability with respect to the physical demands of your work? The values are compared according to store size.

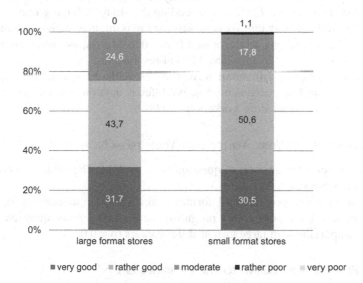

FIGURE 8.3   How do you rate your current work ability with respect to the mental demands of your work? The values are compared according to store footprint.

In total, as many as 75.4% of surveyed women working in large format stores and 81.1% of women working at small format stores rated their capacity vs. the required mental effort as *very good* or *fairly good*.

No statistically significant differences were found in the scores corresponding to question no. 2 of WAI, related to mental effort assessment in workers. The values are compared according to store size.

THE NUMBER OF CURRENT DISEASES DIAGNOSED BY A PHYSICIAN AND PATIENTS

Statistically significant higher values corresponding to question no. 3 of WAI (fewer diseases) were obtained by women working in small format stores as compared to those employed in large format stores (p = 0.000).

THE ESTIMATED WORK IMPAIRMENT DUE TO DISEASES

More than half of the women surveyed –51.3% did not report diseases that would in any way impair their ability to work or affect the quality of work performed, another 48.3% of women surveyed believed that the diseases or injuries have a greater or lesser impact on scope and possibilities of providing work and 0.3% of respondents were of the opinion that they are completely unable to work.

The women employed in small format stores obtained statistically significant higher scores corresponding to question no. 4 of WAI (fewer disease-related work ability restrictions) in comparison with large format store workers (p = 0.000).

SICK LEAVE DURING THE PAST YEAR (12 MONTHS)

In the surveyed group of women, the majority of female respondents –41.3% did not take any sick leave in the 12 months preceding the study. A large group of respondents –33.3% did not work due to dismissals for a maximum of 9 days. Absence from work due to health reasons, which ranged from 10 to 24 days, was reported by 21.3% of respondents, and over 25 days – by 4.0% of respondents.

The women employed in small format stores achieved statistically significant higher scores regarding question no. 5 of WAI (fewer days on sick leave) in comparison with the large format store workers (p = 0.000).

OWN PROGNOSIS OF WORK ABILITY TWO YEARS FROM NOW

Figure 8.4 presents the answers to question no. 6 of WAI. The values are compared according to store size.

The women employed in small format stores achieved statistically significant higher scores, to their work ability prognosis for the next 2 years (question no. 6 of WAI) in comparison with large format store workers (p = 0.003).

MENTAL RESOURCES REFERRING TO THE WORKER'S LIFE IN GENERAL, BOTH AT WORK AND DURING LEISURE TIME

The score is assessed based on the answers to three sub-questions, related to such aspects as *satisfaction with doing everyday activities, being active and spry* and *feeling hopeful for the future*. The total score corresponding to answers to this question may vary from 1 to 4, where 1 means *very small mental resources* and 4 *big mental resources*.

Figure 8.5 shows the answers to questions included in subsection no. 1, regarding question no. 7 of WAI, according to store footprint.

Out of the surveyed women, 21.3% reported being always active and dynamic, whereas in 40.7%, the answer was *fairly often*. The answers indicating poor activity,

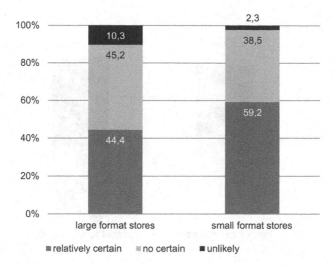

FIGURE 8.4   Do you believe that – from the standpoint of your health – you will be able to do your current job two years from now? The values are compared according to store size.

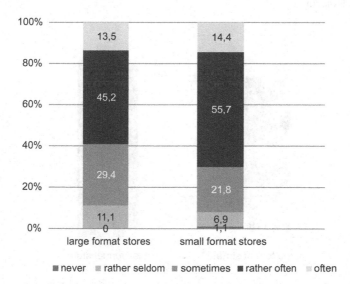

FIGURE 8.5   Have you recently been able to enjoy your regular daily activities? The values are presented according to shop size.

constituted merely 9,3% of all the answers and 1% of surveyed women was entirely inactive in the recent time.

Figure 8.6 shows the scores corresponding to answers to subsection no. 2 regarding question no. 7 of WAI. The values are presented according to store size.

Figure 8.7 shows the answers to subsection no. 3, regarding question no. 7 of WAI, taking into account store size.

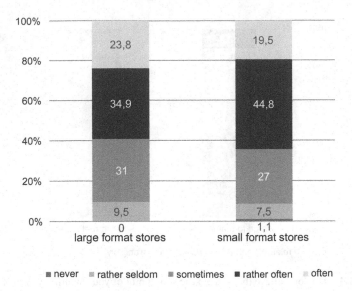

FIGURE 8.6 Have you recently been active and alert? The values are compared according to the store footprint.

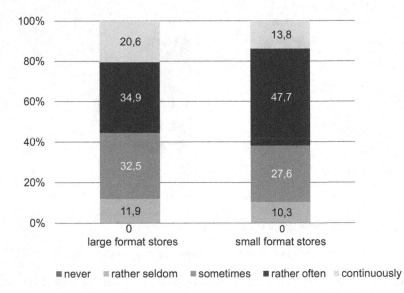

FIGURE 8.7 Have you recently felt yourself to be full of hope for the future? The values are compared according to the store footprint.

DISCUSSION

Hard physical work belongs to the occupational activities that lead to significant age- and gender-related limitations in job performance. Meanwhile, the research on working conditions in European countries has found that the requirements placed on employees remain unchanged, regardless of their age (Bethge et al. 2012). This finding indicates that the actual work burden is increasing alongside the worker's age. Excessive physical load applied to workers regardless their age and gender as well as such factors as health status or motor ability contributes to work ability deterioration and, in consequence, worsens the quality and efficiency of job performance. Besides, it may result in an increased number of accidents at work and absenteeism due to various conditions and diseases.

The primary endpoint of the study was to assess work ability among women aged 55 years and above, involved in physical work in the retail sales sector, including commercial sites of various selling areas. A survey was carried out in a group of 300 women who met the following criteria: aged 55 years or above, occupationally active, employed at the retail trade sector, doing jobs involving manual labour and with at least 5 years of work experience in the retail trade sector. The questionnaire designed and developed for the survey included questions, concerning personal data, occupational activity and ability to work in the surveyed group of women. As the range and type of the work performed by retail sector workers depends mostly on the size of a retail unit, the women selected for the research worked in both large format stores (hypermarkets, supermarkets) and small format stores (local ones). The women employed in large format stores worked more hours a week, more often at night and in evenings as compared to those working in small format stores, and the differences were statistically significant. The values reflecting a longer exposure to lifting and/or handling of heavy loads, pushing and/or pulling heavy loads using assistive devices, or corresponding to repetitive work were also statistically significant. There are few studies assessing the level and type of burdens in the retail trade in workplaces where physical (manual) labour is required. One of such studies, targeting this particular group of employees, was carried out by researchers at the Department of Work Physiology and Ergonomics, Nofer Institute of Occupational Medicine in Lodz. This research project targeted a group of persons, representing all workplaces at trade sites (Bortkiewicz et al. 2011). The sample selected for the study included salesmen, personnel of manufacturing departments or the workers responsible for preparing products for sale, packagers, cashiers, warehousemen and others. In total, workload was assessed at 31 workplaces with female workforce and at 42 workplaces with male workforce. The research findings indicate that the issue of musculoskeletal pain affecting the staff of hyper- and supermarkets requires attention from both employers and physicians who provide prophylactic care to this occupational group. The fact that as much as 83% of the employed women reported various complaints, with 63% reporting frequent symptoms (with weekly or higher incidence), is very concerning, especially as the studied group included relatively young women (their mean age was 30 years). It may be expected that the incidence and severity of complaints will continue to increase with age.

The main goal of the study was to determine work ability levels in the studied group of female workers.

Work ability is a complex feature. Its level is a result of interaction between job requirements concerning physical and mental effort capacity, the worker's functional potential, health status and self-rating of functioning under specific organizational and social circumstances. In this sense, work ability reflects well-balanced requirements of the job and the worker's potential.

The mean WAI value obtained for the whole studied group of women was 38.52, ranking it in the category of good ability to work. Work ability was statistically significantly higher in women aged 55–60 years as compared to the female workers aged 60 years or above. The size of a retail unit and the assortment sold in a retail unit were the values significantly differentiating WAI values in the studied sample. The work ability of women employed in small format stores was better as compared to that of the women employed in large format stores. The lower work ability of the latter resulted from both objective (more diagnosed diseases and conditions, more days of absence due to health reasons) and subjective factors (more limitations in work performance due to health reasons and worse prognoses for their work ability within the next 2 years). One of the basic conditions for a long-lasting good work ability is a proper balance between job requirements and the worker's individual potential. Unfortunately, as research shows, professional life seems to have a varied impact of ageing process (Ilmarinen 2006). The lower levels of work ability in women employed in large format stores possibly resulted from job requirements that were disproportionate to their physical capacity and difficulties in adjusting changes at work to changes in their individual resources. The values obtained from the survey indicate that a greater workload was imposed on women employed in large format stores, due to a higher energy expenditure while carrying, pushing or pulling heavy loads. Their working days were also found to be longer, which possibly resulted from a too short and ineffective rest after work.

The WAI was used in numerous prospective studies to trace the age-related changes in work ability (Bonsdorff et al. 2011; Bischman et al. 2014). The goal of the study by Bugajska was to explore the effect of physical capacity on the ability to work among occupationally active men (n = 664) and women (n = 524) in Poland (Bugajska et al. 2005). Regarding the studied women, the highest WAI values were noted in the age group of 25–30 years, whereas the lowest ones were found in the participants aged 50 years and above. As regards the male participants, the highest WAI scores were obtained in young men (18–24 years old), whereas the lowest values were noted in the group aged above 50 years, as in case of the women from this age group. WAI has also been applied to study the differences in work ability, related to the kind of job and the type and magnitude of workload at workplace. The study has found that the age-related pace of work ability decrease depended on such internal factors as health status as well as external factors associated with the performed job. Research results suggest that such factors as high energy expenditure at work, repeatable activities, a forced, uncomfortable body position and prolonged work time are the significant independent factors responsible for a premature complete loss of work ability. A study carried out at the Nofer Institute of Occupational Medicine in Lodz examined the effects of occupational and extra-professional (life style) factors on work ability. The sample included 1,194 workers of both sexes, aged 18–63 years (Makowiec-Dąbrowska et al. 2008). The results demonstrated the age-related decline in work ability. In the case of women, their self-evaluation of psychological resources improved with age, while the prognosis of work ability for the next 2 years did not change. All other elements of work ability were found to deteriorate with age.

While tracing the age-related changes in work ability, attention was focused on possible modifications of this parameter. It was found that the worker's subjective assessment of work ability could be improved by physical load reduction and an increased participation in leisure time activities (Tuomi et al. 1997).

To sum up, the results of the survey indicate that the sales area and the related working arrangements were the factors differentiating a statistically significant load impact of individual occupational factors and work ability in the studied sample of women. Large format store workers were found to be more burdened by the main job-related physical activities such as lifting, pushing and pulling, as compared to small format store workers. Their work ability also turned out to be worse compared with that of small store staff.

It seems that some solutions involving workload optimization and adaptation of working environment should be proposed to improve work ability in super- and hypermarket workers. Healthy lifestyle promotion, as well as prophylactic measures, focused mainly on physical capacity improvement in this group of workers is also advised. These actions, taken at the right moment of the worker's professional life, allow for maintaining good health, continuing work in satisfactory conditions and preventing a premature exclusion from the group of professionally active people.

REFERENCES

Andrzejczak, D., A. Mikina, B. Rzeźnik, and M. Wajgner. 2010. *Podstawy działalności handlowej [Basics of commercial activity]*. Warszawa: Wydawnictwa Szkolne i Pedagogiczne.

Augustyniuk, K., A. Leśniak, M. Szkup, D. Rogalska, and A. Jurczak. 2016. Występowanie dolegliwości bólowych kręgosłupa wśród pracowników sektora handlowego [The occurrence of spinal pain in trade sector workers]. *Forum Medycyny Rodzinnej* 10(2):95–98.

Bethge, M., F. Radoschewski, and C. Gutenbrunner. 2012. The work ability index as a screening tool to identify the need for rehabilitation: longitudinal findings from the second German sociomedical panel of employees. *J Rehabil Med* 44(11):980–987.

Bischman, J., H. Van der Molen, M. Frings-Dresen, and J. Sluiter. 2014. The impact of common mental disorders on work ability in mentally and physically demanding construction work. *Int Arch Occup Environ Health* 87(1):51–59.

Bonsdorff, A., J. Seitsamo, J. Ilmarinen, C. H. Nygard, M. E. Bonsdorff, and T. Rantanen. 2011. Work ability in midlife as a predictor of mortality and disability in later life: a 28-year prospective follow-up study. *CMAJ* 183(4):E235–E242. DOI: 10.1503/cmaj.100713.

Bortkiewicz, A., T. Makowiec-Dąbrowska, W. Koszada-Włodarczyk, and Z. Jóźwiak. 2011. Obciążenie pracą i związane z tym dolegliwości u pracowników super- i hipermarketów. [Workload and related complaints in super- and hypermarket workers]. *Praca i Zdrowie* 11:38–40.

Bugajska, J., T. Makowiec-Dąbrowska, A. Jegier, and A. Marszałek. 2005. Physical work capacity (VO$_2$max) and work ability (WAI) of active employees (men and women) in Poland. *Int Congr Ser* 1280:156–160. DOI: 10.1016/j.ics.2005.03.001.

GUS. 2019a. *Osoby powyżej 50. roku życia na rynku pracy w 2018. [People over 50 on the labor market in 2018]*. Warszawa: Główny Urząd Statystyczny.

GUS. 2019b. *Rynek wewnętrzny w 2018*. [*Internal market in 2018*]. Warszawa: Główny Urząd Statystyczny.

Ilmarinen, J. 2006. *Towards a longer worklife. Ageing and the quality of worklife in the European Union*. Helsinki: Finnish Institute of Occupational Health, Ministry of Social Affairs and Health.

Kierunki. 2014. Made in Poland [Trends 2014. Made in Poland]. Raport DNB Poland and Deloitte. https://www.dnb.pl/pl/kierunki/kierunki-2014/

Koszada-Włodarczyk, W., Z. Radwan-Włodarczyk, and T. Makowiec-Dąbrowska. 2001. Koniec ciężkiej pracy fizycznej? [Is it the end of the hard work?]. *Atest* 11:4–5.

Kozłowski, S., and K. Nazar. 1995. *Wprowadzenie do fizjologii klinicznej. [Introduction to clinical physiology]*. Warszawa: Wydawnictwo Lekarskie PZWL.

Makowiec-Dąbrowska, T., W. Koszada-Włodarczyk, A. Bortkiewicz et al. 2008. Zawodowe i pozazawodowe determinanty zdolności do pracy. [Occupational and non-occupational determinants of work ability]. *Med Pr* 59(1):9–24.

Momot, R. 2016. *Rynek detalicznego handlu spożywczego w Polsce [The market of foodstuff trade in Poland]. Raport Fundacji Republikańskiej. [Report of the republican foundation]*. Warszawa: Fundacja Republikańska. http://www.pih.org.pl/images/dokumenty/FR_Raport-Rynek-detalicznegi-handlu-spoywczego.pdf. (accessed September 18, 2019).

Pasternak, J., and J. Musiał. 2008. *Rodzaje sklepów. [Types of shops]*. Encyklopedia zarządzania. https://mfiles.pl/pl/index.php/Rodzaje_sklep%C3%B3w (accessed September 18, 2019).

The 6th European Survey of Working Conditions. 2015. The European Foundation for the Improvement of Working and Living Conditions. https://www.eurofound.europa.eu/pl/surveys/2015/ewcs-2015-research-reports (accessed September 18, 2019).

Tuomi, K., J. Ilmarinen, A. Jahkole, L. Karajaninne, and A. Tulkki. 1998. *Work ability index*. 2nd revised ed. Helsinki: Finnish Institute of Occupational Health.

Tuomi, K., J. Ilmarinen, R. Martikainen, L. Aalto, and M. Klockars. 1997. Aging, work, life-style and work ability among Finnish municipal workers in 1981–1992. *Scand J Work Environ Health* 23(Suppl 1):58–65.

9 Work Ability among Male Workers in Different Ages – Results of Research

Marzena Malińska
Central Institute for Labour
Protection – National Research Institute

CONTENTS

INTRODUCTION

The conception of Work Ability Index (WAI) was developed 35 years ago at the Finnish Institute of Occupational Health (FIOH), and it is now a widely used tool of subjective work ability assessment. Work ability is defined as the balance between the worker, his/her resources, health status and skills and work demands. According to the authors of this conception, the strongest relationship is observed between man's health (both physical and mental) and working conditions (work organization, relationships with coworkers, management and leadership) (Ilmarinen and Lehtinenand 2004; Ilmarinen 2007). The influence of the worker's family, friends and the closest environment on his/her work ability, the new trends in the world's economy related to globalization and the use of new technologies or digitalization, together with the present demographic transition, are equally important factors (Ilmarinen et al. 2015). Such perception is a reliable predictor of work ability levels in workers, and in the context of the progressing demographic transition, it seems particularly important. This especially concerns Poland, where very low values of employment activity rate are observed in people above 50. In 2017, the average employment activity rate in the Polish population aged 55–64 years was 58.4%, so it was 22% lower than the corresponding value obtained for the entire population. At the same time, the corresponding mean value of this parameter in the EU was 63.9% (MRPiPS 2017). In addition, the Labor Force Survey Data from 2015 show that only every third person aged 50+ was occupationally active that time (GUS 2016).

WAI is a simple and quick tool, having numerous advantages from the point of view of both research and practice. It is a questionnaire, developed to support workers and used to ensure (especially in an early phase of work) that all the intervention measures have been applied to maintain one's work ability level. The questionnaire consists of seven items, including two items which are objective indicators of health status based on a number of current diseases or injuries and days on sick leave during the past year. The remaining items, in turn, pertain to subjective parameters reflecting work ability, such as (1) current work ability as compared with lifetime best (top form), (2) work ability compared with physical demands of work (physical and mental effort), (3) estimation of disease-related work ability impairment and (4) own prognosis of work ability two years from now and mental resources (including work satisfaction, optimism and enthusiasm) (Ahlstrom et al. 2010). Work ability is determined by multiple variables. Assessment of the factors responsible for a decline in work ability is an important process, aimed at prevention of premature impairment of work ability and the development of more accurate recommendations, as part of health-promoting programs.

The presented research aimed at subjective assessment of the perceived work ability in men and determining the effect of selected occupational and non-occupational factors on chances to maintain high work ability levels in the male population, irrespective of their age. The sample assessed in the above-mentioned study included only male workers. Although statistics show that life expectancy in men is 8 years shorter than in women, the period of occupational activity is longer in the male population. Research results also show that men more often have accidents at work and work under conditions which pose a risk for their health. Men suffer from cardiovascular diseases and cancer more often than women. In addition, the average education level is lower in the male population as compared with the female population, and men less care about their health than women and more often lead an unhealthy lifestyle. The long-term data on alcohol consumption reveal a higher percentage of alcohol addicts among the male population in Poland, even higher than among female tobacco smokers, and a higher percentage of overweight and obese men (GUS 2016; Czapiński and Panek 2015; Cianciara 2012).

MATERIALS AND METHODS

Work ability was assessed using the subjective WAI (Tuomi et al. 1998).

The questionnaire pertaining to workers' lifestyle was developed for this research. It included questions about age, education level, the overall work experience and work experience in their current job, work position and working system, as well as body height and body mass for body mass index (BMI) calculation. Next items of the questionnaire included self-rating of health status and nutrition, tobacco smoking, and alcohol and coffee consumption.

Psychosocial and physical work demands were assessed using Job Content Questionnaire (JCQ) developed by Robert Karasek (1979). The questionnaire measures three variables determining stress levels at workplace: Psychological Demands (nine items), "Decision Latitude" (nine items) and "Social Support" (eight items). For the purpose of this study, "Physical Exertion" (five items) and "Hazard at Work" (eight items) scales were used, based on the questions about physical effort, exposure to dangerous and toxic factors as well as overall exposure to physical factors.

The assessment of musculoskeletal symptoms was performed using Nordic Musculoskeletal Questionnaire (Kuorinka et al. 1987). The questionnaire included items assessing the prevalence of symptoms and limitations in usual activity performance due to such symptoms affecting nine parts of the musculoskeletal system within the last 7 days and 12 months.

The survey was conducted in a sample of 2000 men. The participants were selected using quota and judgmental sampling. The inclusion criteria comprised age and type of job.

The sample included men of different ages, performing different types of jobs, which is presented as follows:

- 20–25 years (200 physical workers and 200 mental workers)
- 30–35 years (200 physical workers and 200 mental workers)
- 40–45 years (200 physical workers and 200 mental workers)
- 50–55 years (200 physical workers and 200 mental workers)
- 60+ (200 physical workers and 200 mental workers).

RESULTS

A sample of men performing mental (n = 1000) and physical work (n = 1000) aged 20–65 years (42.1 ± 14.1 years) participated in the study. The sample included 53% and 26.3% of men with higher and secondary education, respectively, and 11% of men with vocational education. The overall mean working experience in the studied sample was 19.5 years (SD = 13.2), and their mean working experience in current jobs was 8 years and 3 months (SD = 7.7). The majority of study participants worked in the following sectors: construction (33.8%), trade and services (25.3%), public administration (8.6%), banking and finance (7.9%) and IT (7.1%). The participants worked on average 43.6 hours a week (SD = 7.7). In almost 45% of participants, the body mass was normal (BMI 18.5–24.99), 43.2% were overweight (BMI 25–29.99), and over 7% were obese (30–34.99).

Most of the studied men reported an excellent (42.7%) and good (37.6%) work ability. A moderate work ability was noted in 15.7%, whereas 1.4% of the participants reported poor work ability. The mean work ability value in the entire sample was 41.6 (SD = 5.4). Analysis of the results revealed age-related decline in work ability. In the group of men aged 20–25 and 30–35 years, work ability was significantly more often rated as excellent as compared with that of men aged 50–55 and 60 years and above. In the latter group, a higher percentage of men with poor work ability was noted (Table 9.1.). The results showed a statistically significantly higher percentage of men with higher education, holding a managerial position. A higher percentage of men reporting poor work ability, in turn, was noted among physical workers (WAI 40.6 ± 5.6) with long working experience (above 30 and 40 years) and overweight, as compared to those performing mental work (WAI 42.6 ± 5.0) with normal body mass

TABLE 9.1

Work Ability Index in Different Age Groups

	20–25 Years	30–35 Years	40–45 Years	50–55 Years	60+	Total
Total WAI value; mean value ± SD)	44.5 ± 3.8[a]	43.3 ± 4.3[a]	42.2 ± 4.4[a]	39.4 ± 5.7[a]	38.4 ± 5.9[a]	41.6 ± 5.4
Work ability rating [%participants]						
Poor (7–27 points)	0.0	7.1	0.0	35.7[b]	57.1[b]	1.4
Moderate (28–36 points)	4.5	8.3	16.2	33.4	37.6	15.7
Good (37–43 points)	15.7	18.5	21.7	21.3	22.9	37.6
Excellent (44–49 points)	31.1[b]	26.3[b]	20.5	13.4	8.8	42.7

[a] Statistically significant differences between the participants aged 20–25 and 30–35 years, 40–45, 50–55 and 60 years and above; and the participants aged 30–35 years, 40–45, 50–55 and 60 years and above (p ≤ 0.05).

[b] Statistically significant differences between the participants aged 20–25 and 30–35 years, and the participants aged 50–55 years and 60 years and above.

and work experience up to 5 years. Among the men working in one-shift system, the highest percentage of those reporting an excellent work ability was noted. The highest percentage of participants rating their work ability as poor, in turn, was noted among those working in three-shift and flexible working hour systems (p ≤ 0.05) (Table 9.2.).

The presented data indicate that the men in the following age groups – 20–25, 30–35 and 40–45 years – reported higher levels of work ability as compared with lifetime best, their ability to overcome difficulties related to mental and physical work performance and being full of hope for the future (p ≤ 0.05). The oldest participants, in turn, significantly more often were on sick leave and reported higher levels of work ability impairment due to health problem (p ≤ 0.05) (Table 9.3.).

A chance for good work ability, measured using WAI, was assessed. Multiple regression analysis was carried out for this purpose, and WAI was divided into two levels: low (0) related to poor and moderate work ability rating; and high (1) related to good and excellent work ability. The results of the above-mentioned analyses revealed multiple age-related factors contributing to high, statistically significant levels of work ability (Table 9.4). Among the individual traits having a negative impact on work ability levels, poor self-assessment of health (in males aged 20–25 years and 60 years and above) and dietary habits (in men aged 30–35 years, 40–45 years and 50–55 years), smoking more than 14 cigarettes a day (in men aged 40–45 years) as well as symptoms of musculoskeletal disorders with a negative impact on work ability levels were found in all groups, except that of the youngest men. The most frequently noted complaints affecting WAI included neck pain (men aged 30–35 years, 50–55 years and 60 years and above), low back pain (men aged 40–45 years, 50–55 years and 60 years and above) and hip pain (men aged 30–35 years and 40–45 years). Drinking coffee was the factor increasing the chance for high WAI levels (men aged 30–35 years and 40–45 years).

TABLE 9.2
Work Ability Assessment by Age, Education Level, Total Work Experience, Position, Work System and BMI Values [% Participants]

	Work Ability WAI			
	Poor (≤27 Points)	Moderate (28–36 Points)	Good (37–43 Points)	Excellent (44–49 Points)
Education level:				
Primary/basic	12.0	13.5	8.6	6.6
Vocational	28.0[a]	18.4	12.6	6.3
Secondary	52.0	43.5	56.6	55.7
High	8.0	24.5	22.1	31.3[a]
Total work experience				
up to 5 years	0.0	1.5	17.2	33.1[a]
6–10 years	3.6	5.2	9.3	17.7[a]
11–20 years	7.1	9.4	20.3	20.2
21–30 years	7.1	34.8	24.3	17.2
31–40 years	67.9[a]	35.8	24.1	10.0
>40 years	14.3[a]	10.3	6.8	1.8
Type of work				
Mental workers	28.6	39.2	43.4	59.1[a]
Physical workers	71.4[a]	60.8	56.6	40.9
Job position:				
Rank and file	90.0[a]	81.7	84.0	83.0
Managerial	10.0	18.3[a]	16.0	17.0
Working system:				
1 shift	67.9	75.8	79.4	85.1[a]
2 shifts	10.7	12.7[a]	7.9	7.0
3 shifts	7.1[a]	7.0	5.1	1.9
Flexible working hours	14.3[a]	4.5	7.6	6.0
BMI				
Normal body mass	28.0	33.7	43.7	55.3[a]
(BMI 18.5–24.9)	56.0[a]	50.0	48.3	40.6
Overweight (BMI 25–29.9)	16.0	16.3	8.1	4.1
Obesity (BMI > 30)				

[a] Statistically significant differences between the participants: 1/with primary education and high education; 3/managerial and rank and file; 4/working in one-, two- and three-shift systems and flexible hours; 5/with normal body mass and overweight.

The factors responsible for lowered chances to reach high WAI levels included working in big teams (men aged 20–25 years), air pollution in the working environment (in men aged 30–35 years), an awkward body position at work (men aged 40–45 years), intense physical effort at work and lack of control over work performance (men aged 50–55 years) as well as high job insecurity and high work demands (in the group aged 60 years and above) (Table 9.4).

TABLE 9.3
Selected Elements of WAI in the Studied Sample of Males

WAI Elements	a 20–25 Years	b 30–35 Years	c 40–45 Years	d 50–55 Years	e 60+	Total
1. Current work ability as compared to lifetime best [mean value, SD]	8.75 ± 1.29*b, c, d, e	8.21 ± 1.33*d, e	8.04 ± 1.34*d, e	7.31 ± 1.69	7.24 ± 1.84	7.92 ± 1.61
2. Work ability in relation to occupation requirements [mean value, SD]:						
Abilities to manage physical efforts	4.41 ± 0.65*b, c, d, e	4.23 ± 0.74*d, e	4.17 ± 0.63*d, e	3.95 ± 0.83	3.91 ± 0.81	4.14 ± 0.76
Abilities to manage mental efforts	4.45 ± 0.71*d, e	4.37 ± 0.75*d, e	4.34 ± 0.70*d, e	4.11 ± 0.80	4.15 ± 0.76	4.29 ± 0.75
3. Assessment of disease-related impairment of work ability [mean, SD]	5.79 ± 0.53*c, d, e	5.73 ± 0.58*c, d, e	5.54 ± 0.70*d, e	5.25 ± 0.89	5.08 ± 0.94	5.48 ± 0.79
4. Sickness absenteeism during from past 12 months [% workers]:						
100–365 days	0.0	0.5	0.3	1.3	0.0	0.4
25–99 days	1.0	0.5	1.5	1.5	2.4	1.4
10–24 days	4.8	9.7	15.4	17.5	26.2	14.3
At most 9 days	21.4	32.2	29.0	28.5	30.2	27.5
No. of sick leave	72.8	57.0	53.7	51.2	41.2	53.9
5. Self-assessed prognosis of work ability for the coming 2 years [mean value, SD]	6.19 ± 1.57*d, e	6.17 ± 1.64*d, e	6.38 ± 1.31*d, e	5.35 ± 1.98	5.30 ± 1.98	5.88 ± 1.77
1. Mental resources of work performance [mean value, SD]	3.25 ± 0.74*c, d, e	3.16 ± 0.78*d, e	3.10 ± 0.71*d, e	2.82 ± 0.83	2.77 ± 0.82	3.02 ± 0.80
Ability of enjoying work	3.24 ± 0.80*d, e	3.09 ± 0.87*d, e	3.09 ± 0.85*d, e	2.73 ± 0.92	2.79 ± 0.95	2.99 ± 0.90
Being active and alert	3.26 ± 0.79*d, e	3.08 ± 0.86*d, e	3.14 ± 0.80*d, e	2.79 ± 0.89	2.85 ± 0.97	3.03 ± 0.88
Being full of hope for the future						

* Statistically significant differences between the participants of different age 20–25 years (a), 30–35 years (b), 40–45 years (c), 50–55 years (d) and 60 years and above (e) (p ≤ 0.05).

TABLE. 9.4

The Results of Multivariate Logistic Regression Analysis Corresponding to Work Ability (WAI) in Men Aged 20–25 Years, 30–35 Years, 40–45 Years, 50–55 Years and 60 years and above (p ≤ 0.05)

Age	Variables	p	Rating	Logistic Regression Model	
				OR	95%CL
20–25 years	Self-rated health	0.0002	−1.474	0.23	0.10–0.51
	Working in big teams	0.0000	−0.115	0.89	0.84–0.95
30–35 years	Hip pain	0.0065	−2.079	0.13	0.03–0.56
	Neck pain	0.0001	−1.790	0.17	0.07–0.41
	Pollution at workplace	0.0002	−1.268	0.28	0.14–0.55
	Self-rated nutrition	0.0039	−0.645	0.52	0.34–0.81
	Coffee consumption	0.0103	1.253	3.50	1.34–9.15
40–45 years	Hip pain	0.0026	−2.415	0.09	0.02–0.43
	Smoking more than 14 cigarettes daily	0.0000	−1.740	0.18	0.08–0.40
	Coffee consumption	0.0126	1.356	3.88	1.33–11.31
	Awkward posture at work	0.0000	−1.231	0.29	0.17–0.50
	Low back pain	0.0100	−1.080	0.34	0.15–0.78
	Self-rated nutrition	0.0000	−0.935	0.39	0.26–0.59
50–55	Low back pain	0.0002	−1.176	0.31	0.17–0.58
	Neck pain	0.0037	−1.026	0.36	0.18–0.72
	Self-rated nutrition	0.0013	−0.506	0.60	0.44–0.82
	Physical effort	0.0361	−0.306	0.74	0.55–0.98
	Control	0.0059	0.040	1.04	1.01–1.07
60+	Self-rated health	0.0000	−1.582	0.21	0.13–0.33
	Low back pain	0.0020	−1.014	0.36	0.19–0.69
	Thoracic pain	0.0050	−0.976	0.38	0.19–0.75
	Neck pain	0.0278	−0.810	0.45	0.22–0.92
	Job insecurity	0.0000	−0.380	0.68	0.58–0.80
	Requirements	0.0223	−0.073	0.93	0.87–1.0

DISCUSSION

Work ability assessment was carried out in a large sample of the male working age population. The mean value obtained in the sample, reflecting subjective perception of work ability, allows rating its levels as good. It is justified for several reasons. First of all, the sample included occupationally active individuals whose mean age was 42 years. Comparing the data obtained in this study with those reported by other researchers, we can see that men in Poland are characterized by the highest levels of work ability. The values are slightly higher than the reference values developed by the Finnish Institute of Occupational Health, similar to Dutch workers' results, but slightly lower than those obtained in a sample of Brazilian workers (Strijk et al. 2011; Monteiro et al. 2006). The values similar to these obtained from our study were also reported by Portuguese researchers who assessed work ability in computer operators (n = 50). The results of data analysis conducted in that group indicate that 78% of workers reported good and excellent work ability, whereas only 2% reported poor work ability. The mean value of WAI in that sample was 40.5 (SD = 5.76) (Costa et al. 2011).

Our results are more optimistic than the data presented in 2008 by a team of other Polish researchers. In their study, the mean value of WAI in the male sample (n = 669) was 1.7 points lower than in our study. The study results also show that the highest work ability levels were found in men aged 30–39 years, including 33.5% with excellent work ability, 56.7% with good work ability and 9.8% with moderate work ability. Interestingly, the highest WAI values were noted in males exposed mainly to mixed workload (WAI 41.27) and physical workload (WAI 39.31), whereas the lowest scores were found in mental workers (WAI 38.76) (Makowiec-Dąbrowska et al. 2008).

The results of our study, in turn, indicate that the highest scores were noted in the group of mental workers (42.6 WAI) with a statistically higher percentage of men reporting excellent work ability as compared to the group of physical workers with the highest percentage of men with poor work ability.

Camarino in the study being part of NEXT project between the years 2002 and 2004, assessing work ability levels in the female and male nursing personnel in Europe (including 14.2% of men, n = 1063), found that the lowest values of WAI were noted in Polish health service workers, with the mean value of 37.6 ± 5.8. The highest WAI values were obtained in the Dutch (41.8 ± 4.8), Italian (39.8 ± 5.3), Belgian (39.4 ± 4.8) and French (38.6 ± 5.3) workers' populations (Camerino et al. 2008).

As numerous literature data show work ability significantly decreases with age in the male population (Kiss et al. 2002; Monteiro et al. 2006; Lin et al. 2006; Firoozeh et al. 2017; Labbafinejad et al. 2014; Karttunen and Rautiainen 2009; Strijk et al. 2011; Makowiec-Dąbrowska et al. 2008; Bugajska et al. 2008; Marszałek et al. 2005). The results of our study reveal such correlations. The workers aged 20–25 and 30–35 years were characterized by excellent work ability as compared with the group of men aged 50–55 years and 60 years and above where the highest percentage of poor work ability was observed.

The results obtained from the large-scale research carried out by Kujala in a sample of 2021 men and 1704 women indicate very high values of WAI in young workers (mean age = 31 years ± 0.5). The mean WAI value obtained from data analysis is 41.1 (SD = 4.2) in young men, and slightly lower values are noted in women 40.1 (SD = 4.2) (Kujala et al. 2005). Likewise, the results obtained by Lin in a large sample of Chinese men (n = 4405) indicate a statistically significant age-related decline in WAI values. The highest WAI values are noted in young persons below 29 years (WAI 39.9 ± 5.3), and the lowest ones are found in persons aged 40–49 (37.6 ± 6.4) and after 50%. (35.8 ± 3). The highest percentage of participants with excellent work ability is noted in men between 30 and 39 years (21%), and up to 29 years (20%). The highest percentage of men with poor work ability is noted in those aged 40–49 years (8%) and 50 years and above (13%) (Lin et al. 2006).

Furthermore, the results obtained by Monteiro in a sample of 236 men (49 ± 10.7 years) indicate that the highest values of WAI were noted among persons aged

20–34 years (WAI 43.6), whereas the lowest scores were obtained in persons aged 55–69 years (WAI 40). Logistic regression analysis of these data indicates that the factors related to the risk of work ability decline apart from ageing, lower education levels and long working experience (Monteiro et al. 2006). Similar correlations were also observed by the authors of WAI who reported age-related decrease in work ability values, and the speed of this decrease depended on health-related factors as well as the type and intensity of work (Tuomi et al. 1991; Tuomi et al. 1997). The study on the effect of lifestyle on work ability is the issue of interest for many researchers. Kaleta in the study on the effect of the lifestyle on work ability has found strong correlations between the two above-mentioned parameters, despite the relatively small sample size (94 men and 93 women). Analysis of these results revealed statistically significant correlations between work ability and such parameters as lifestyle, tobacco smoking, BMI values and physical activity (Kaleta et al. 2006).

The fact that symptoms of musculoskeletal disorders, particularly in the cervical and lumbosacral spine segments, decreased the chance for obtaining high WAI scores is of note. Interestingly, the results presented by Bugajska indicate that, despite such complains reported by both younger (<45 years) and older workers, only in the latter group, they were considered risk factors for a decreased work ability as compared with lifetime best (Bugajska and Sagan 2014).

In the majority of the studied men (aged 30–35, 40–45 and 50–55 years), the quality of diet favourably affected their WAI values. The results of our analysis suggest that a healthy diet (subjectively assessed) increased the chance for obtaining higher mean WAI scores. Similar results were observed by the team of researchers headed by Makowiec-Dąbrowska indicated that a well-balanced diet is a protective factor against work ability impairment. The sample of the above study comprised 1194 workers including 669 men aged 18–63 years (Makowiec-Dąbrowska et al. 2008). The presented data thus confirm that solutions aimed at health promotion at workplace are required, with particular emphasis on prevention of musculoskeletal disorders conditions as well as training and panel meetings dedicated to healthy nutrition in the male population.

Among the variables reflecting a significant negative impact of work on chances for very good work ability, the psychosocial work demands (job insecurity, work demands and control over work tasks), physical factors (related to high physical workload, air pollution and awkward posture) as well as organizational factors, related to work in large teams, have the highest impact of work ability deterioration. Our results seem to be in agreement with the data presented in 2005 by Ilmarinen, particularly these related to the oldest age group. The authors report that the strongest work ability determinants include health status, working conditions, family relations as well as hobbies and interests, and in persons aged 55–64 years, these are usually health status, physical fitness and mental stress (Ilmarinen 2005). The analysis of all the studies conducted after 1985 using WAI by Berg has revealed that the main factors responsible for work ability deterioration included workers' age, physical inactivity, musculoskeletal disorders, obesity, and high psychosocial and physical work demands (van den Berg 2008).

The issue of ageing society has been widely discussed in recent years. The analysis of our study results has shown numerous alarming data, usually obtained in groups of older workers. The values obtained in a sample of Polish men after 50 reflect poor work ability reported by this group. They more seldom reported being full of hope for the future or enjoying their regular daily life activities, their self-rating of current work ability as compared with lifetime best was lower and they were rather sure that they would do their present job within the next two years. Therefore, age should be regarded as the main factor modifying the development of solutions allowing the male workers to maintain good work ability or improve their work ability. In the case of men after 50 years of age, the necessary solutions include health status improvement (especially prevention of musculoskeletal disorders) as well as improvement of psychosocial work demands by limiting such adverse phenomena as job insecurity and high work demands. In the case of the youngest group of study participants, health promotion and determining the optimum team size are of note. In the case of men aged 30 and 40 years, the recommended solutions include health promotion (prevention of musculoskeletal disorders, taking care of health) and healthy lifestyle promotion (workshops and education trainings about healthy nutrition and the adverse effect of tobacco smoking), and improvement of these men's working conditions involving reduction of air pollution at work and avoidance of awkward body posture.

REFERENCES

Ahlstrom, L., A. Grimby-Ekman, M. Hagberg, and L. Dellve. 2010. The work ability index and single-item question: associations with sick leave, symptoms and health – a prospective study of women on long term sick leave. *Scand J Work Environ Health* 36(5):404–412. DOI: 10.1080/10803548.2014.11077069.

Bugajska, J., T. Makowiec-Dąbrowska, and M. Konarska. 2008. *Zapobieganie wcześniejszej niezdolności do pracy: założenia merytoryczne*. Warszawa: CIOP PIB.

Bugajska, J., and A. Sagan. 2014. Chronic musculoskeletal disorders as risk factors for reduced work ability in younger and ageing workers. *Int J Occup Saf Ergon* 20(4):607–615.

Camerino, D., P. M. Conway, S. Sartori et al. 2008. Factors affecting work ability in day and shift-working nurses. *Chronobiol Int* 25(2–3):425–442. DOI: 10.1080/07420520802118236.

Cianciara, D. 2012. Zdrowie – męska rzecz. Raport Siemensa. http://www.siemens.pl/pool/healthcare/raport_siemensa_2012.pdf (accessed September 18, 2019).

Costa, A. F., R. Pugal-Leal, and I. L. Nunes. 2011. An exploratory study of the work ability index (WAI) and its components in a group of computer workers. *Work* 39(4):357–367. DOI: 10.3233/WOR-2011-1186.

Czapiński, J., and T. Panek. 2015. *Diagnoza społeczna 2015. Warunki i Jakość Życia Polaków*. Warszawa: Rada Monitoringu Społecznego. http://diagnoza.com (accessed September 18, 2019).

Firoozeh, M., M. Saremi, A. Kavousi, and A. Maleki. 2017. Demographic and occupational determinants of the work ability of firemen. *Journal of occupational health* 59(1):81–87. DOI: 10.1539/joh.15–0296-FS.

GUS [Główny Urząd Statystyczny]. 2016. Aktywność ekonomiczna ludności Polski w latach 2013–2015. Warszawa [Statistics Poland. 2016. Economic activity of the Polish population in 2013–2015. Warsaw]. http://www.stat.gov.pl (accessed September 18, 2019).

Ilmarinen, J., and S. Lehtinenand, 2004. *Past, Present, and Future of Work Ability: Proceedings of the 1st International Symposium on Work Ability, 5–6 September 2001*. Tampere, Finland: Finnish Institute of Occupational Health.

Ilmarinen, J. 2005. *Towards a longer worklife. Ageing and the quality of worklife in the European Union.* Helsinki: FIOH Bookstore. http://hawai4u.de/UserFiles/ Ilmarinen_2005_Towards%20a%20Longer%20Worklife.pdf (accessed September 18, 2019).

Ilmarinen, J. 2007. The work ability index (WAI). *Occup Med (Lond)* 57(2):160. DOI: 10.1093/occmed/kqm008.

Ilmarinen, V., J. Ilmarinen, P. Huuhtanen, V. Louhevaara, and O. Nasman. 2015. Examining the factorial structure, measurement invariance and convergent and discriminant validity of a novel self-report measure of work ability: workability – personal radar. *Ergonomics* 58(8):1445–1460. DOI: 10.1080/00140139.2015.1005167.

Kaleta, D., T. Makowiec-Dąbrowska, and A. Jegier. 2006. Lifestyle index and work ability. *Int J Occup Med Environ Health* 19(3):170–177.

Karasek, R. 1979. Job demands, job decision latitude, and mental strain: implication for job redesign. *Adm Sci Q* 24(2):285–308.

Karttunen, J. P., and R. H. Rautiainen. 2009. Work Ability Index among Finnish dairy farmers. *Journal of Agricultural Safety and Health* 15(4):353–364.

Kiss, P., M. Walgraeve, and M. Vanhoorne. 2002. Assessment of work ability in aging fire fighters by means of the Work Ability Index Preliminary results. *Arch Public Health* 60:233–243. https://www.wiv-isp.be/Aph/pdf/aphfull60_233_243.pdf (accessed September 18, 2019).

Kujala, V., J. Remes, E. Ek, T. Tammelin, and J. Laitinen. 2005. Classification of Work Ability Index among young employees. *Occup Med* (Lond) 55(5):399–401.

Kuorinka, I., B. Jonsson, A. Kilbom et al. 1987. Standardised Nordic questionnaires for the analysis of musculoskeletal symptoms. *Appl Ergon* 18(3):233–237.

Labbafinejad, Y., M. Ghaffari, B. Bahadori et al. 2014. The effect of sleep disorder on the work ability of workers in a car accessories manufacturing plant. *Med J Islam Repub Iran* 28:111.

Lin, S., Z. Wang, and M. Wang. 2006. Work ability of workers in western China: reference data. *Occup Med (Lond)* 56(2):89–93.

Makowiec-Dąbrowska, T., W. Koszada-Włodarczyk, A. Bortkiewicz et al. 2008. Zawodowe i pozazawodowe determinanty zdolności do pracy [Occupational and non-occupational determinants of work ability]. *Med Pr* 59(1):9–24.

Marszałek, A., M. Konarska, and J. Bugajska. 2005. Assessment of work ability in hot environment of workers of different ages. *Int Congr Ser* 1280:208–213.

MRPiPS [Ministerstwo Rodziny, Pracy i Polityki Społecznej]. 2017. Osoby powyżej 50. roku życia na rynku pracy w latach 2016–2017 [Ministry of Family, Labour and Social Policy. 2017. People over 50 on the labor market in 2016–2017]. https://psz.praca. gov.pl/documents/10828/4803132/Osoby%20powy%C5%BCej%2050%20roku% 20%C5%BCycia%20na%20rynku%20pracy%20w%202016%20roku.pdf/76524193- e831-4f11-a49c-05be7285c5e9?t=1495455478629 (accessed September 18, 2019).

Monteiro, M. S., J. Ilmarinen, and H. R. C. Filho. 2006. Work ability of workers in different age groups in a public health institution in Brazil. *Int J Occup Saf Ergon* 12(4):417–427.

Strijk, J. E., K. I. Proper, M. M. van Stralen et al. 2011. The role of work ability in the relationship between aerobic capacity and sick leave: a mediation analysis. *Occup Environ Med* 68(10):753–758. DOI: 10.1136/oem.2010.057646.

Tuomi, K., L. Eskelinen, J. E. Toikkanen et al. 1991. Work load and individual factors affecting work ability among aging municipal employees. *Scand J Work Environ Health* (Suppl 1):128–134.

Tuomi, K., J. Ilmarinen, J. Seitsamo et al. 1997. Summary of the Finnish research project (1981–1992) to promote the health and work ability of aging workers. *Scand J Work Environ Health* 23(1, Suppl 1):66–71.

Tuomi, K., J. Ilmarinen, I., A. Jahkola, L. Katajarine, and A. Tulkki. 1998. *Work ability index*. Helsinki: Finnish Institute of Occupational Health.

van den Berg, T. I., L. A. Elders, B. C. de Zwart, and A. Burdorf. 2009. The effects of work-related and individual factors on the Work Ability Index: a systematic review. *Occup Environ Med* 66(4):211–220. DOI: 10.1136/oem.2008.039883.

Part III

*Work Ability and
Chronic Diseases*

10 Work Ability and Its Relationship to Sense of Coherence among Workers with Chronic Diseases – Results of Research

Joanna Bugajska and Maria Widerszal-Bazyl
Central Institute for Labour Protection –
National Research Institute

CONTENTS

INTRODUCTION

Along with the observed demographic transition, we can note an increased share of older people in the working population, including those coping with chronic diseases. The presence of such diseases is one of the main reasons for functional deterioration in older people, limiting their social and occupational activities. At the same time, people with chronic diseases highly value work, most often due to financial independence, but also opportunities for social bonds and active participation in social life (Vooij et al. 2018). In literature, only few papers report the effect of individual factors on occupational activity in people with chronic diseases, irrespective of disease advancement level. Usually, the above issue is perceived in two categories, "occupationally active or inactive" and the psychological factors responsible for work ability level are not analysed.

One of the factors, which are worth noting in this context, is the so-called sense of coherence (SOC). This chapter presents research on this specific factor. The term has been put forward by the medical sociologist, Aaron Antonovsky (1987), to define a general life orientation, fostering better health status and, in case of health deterioration, better disease coping skills.

The SOC consists of three elements:

Comprehensibility denotes an understanding of current circumstances and the people around us, but also the feeling that we can anticipate future events and human behavior;

Manageability means that we have certain resources allowing us to meet the requirements related to specific circumstances; it can be our own or other people's resources we can rely on;

Meaningfulness denotes our conviction that certain areas of life are so important that they are worth our involvement, either cognitive or emotional.

The SOC has been the topic of multiple studies which have shown its effect on health. A systematic literature review carried out by Eriksson and Lindström (2006) has revealed a strong correlation between SOC and health perception, especially as regards mental health, irrespective of age, gender, ethnicity or nationality. The next review carried out by the same authors has also shown the effect of SOC on the quality of life (QoL) with the following conclusion: the stronger the SOC, the better the QoL. The effect of SOC on the QoL, indicating that the stronger the SOC, the better the QoL, has also been noted by the same authors in the subsequent literature review. Besides, the authors suggest that SOC is a predictor of good QoL (Eriksson and Lindström 2007). Other studies have confirmed the thesis that a strong SOC is associated with better HR/QoL levels and self-efficacy as well as less catastrophizing in patients with chronic pain. SOC may be an important coping mechanism (strategy) for patients with chronic musculoskeletal pain (Chumbler et al. 2013).

Multiple studies assessing the SOC were conducted in patients with chronic illnesses. Based on the results obtained in a sample of workers with chronic musculoskeletal pain, it was found that patients with high SOC levels are more resistant to pain (Chumbler et al. 2013). Furthermore, the relationship between SOC and the QoL was noted in patients with coronary artery disease (Silarova et al. 2012), after myocardial infarction (Benyamini et al. 2013; Wrześniewski and Włodarczyk 2012) and among nursing home residents (Jueng et al. 2016). The studies found that even when the degree of disease advancement was considered, the persons with higher SOC values reported better QoL. SOC was also studied in relation to job circumstances in the assessment of the effect of work-related stress on the prevalence of type 2 diabetes in Swedish population (Agardh et al. 2003; Eriksson et al. 2013). Despite multiple studies on SOC, there are no papers reporting analyses of the relationship between the SOC and work ability, measured using Work Ability Index (WAI). Hence, one of the goals of our research, carried out in patients with chronic diseases such as arterial hypertension, coronary heart disease, osteoarthritis or diabetes, was to compare the SOC in four groups of patients differing in types of diseases. Another goal of our research was to determine the relationship between the SOC and work ability (measured using WAI).

STUDY GROUP

Six hundred workers including 200 patients with arterial hypertension, 95 patients with coronary heart disease, 105 patients with diabetes and 200 patients with osteo-arthritis participated in the study. The mean value corresponding to age, obtained in the entire sample, was 50.4 (SD 12.2; min – 20 years, max – 78 years). The highest mean value obtained for age (54.3 years) was noted in the group of patients with coronary heart disease (SD 10.1), whereas the lowest corresponding value (47.7 years) was noted in patients with osteoarthritis (SD 12.4).

Fifty-one per cent of the group of patients with hypertension were women. Eighty-seven per cent participants in this group met the criteria of arterial blood pressure equalization according to ESC/ESH 2013 guidelines. Eighty-five per cent of the participants were diagnosed with grade 1 arterial hypertension, whereas 13% and 2% were diagnosed with grade 2 and 3 arterial hypertension respectively. Ninety per cent were undergoing regular pharmacological treatment, and 33% were on the diet recommended by their physician.

Fifty-one per cent of the group of patients with coronary heart disease were women. The majority of the participants underwent regular pharmacological treatment or presented with advanced stage of the disease class I, according to ECS 2013 grading, and there were only four participants with class II.

Fifty-six per cent of participants with diabetes were women, and there were 96% of patients with type 2 diabetes in the sample. Most of the participants received oral antidiabetic medicines, whereas only six of them received insulin. The majority were on a regular antidiabetic diet, whereas ten participants were on a general diet, without observing any rules of antidiabetic diet. Most of the participants regularly performed self-assessment of glycaemia level, and only five did it occasionally. According to the criterion of diabetes equalization, i.e. the per cent of glycated haemoglobin (HbA1c), corresponding to glycaemia levels from the last 3 months, diabetes was equalized in 90% of the participants.

In the group of patients with osteoarthritis, 71% were women. Twenty per cent of the participants regularly received medications, and in 60% of the group, the course of the disease was stable. In as many as 73% of the participants, osteoarthritis affected the spine, usually its lumbosacral segment, whereas in 33% and 32%, it affected upper and lower limb joints, respectively.

METHODS

Work ability assessment was performed using WAI questionnaire (Tuomi et al. 1998). Cronbach's α value corresponding to WAI was *0.77*. The SOC defined by Antonovsky (1987) as a general life orientation including sense of resourcefulness, comprehensibility and sense was measured using SOC-29 questionnaire. The questionnaire consists of 29 questions, and the answers are measured using 7-point scales with verbal description of extrema. SOS values are calculated based on the total answers to all the questions. The results may range from 29 to 203. In this study, only a global SOC index was assessed. High correlations between the components of SOC are also found in Polish studies (Pasikowski 2001). Also in this study, SOC-29 questionnaire was characterized by a high coherence: Cronbach's α value was *0.89*.

RESULTS

The obtained data revealed the highest WAI values in the group of persons with arterial hypertension (*good work ability* category), whereas the lowest corresponding value was noted in the group with coronary heart disease (*moderate work ability* category). (Table 10.1).

Table 10.2 presents the mean values obtained for SOC levels in four studied groups. The average level of this parameter was higher in groups with arterial hypertension and osteoarthritis and lower in groups of workers with coronary heart disease and diabetes. The differences between the results obtained in groups with arterial hypertension and coronary heart disease, and between the groups with osteoarthritis and diabetes, were statistically significant (Kruskal–Wallis test). Higher SOC values were also found in males and young persons with chronic diseases (Tables 10.3 and 10.4).

Table 10.5 present correlations between SOC and work ability according to WAI. In all the respondents – regardless of the type of chronic disease – the correlation between SOC and the total WAI value, measured using Pearson r, was 0.48 at $p < 0.000$ (the value is not presented in the table). The value obtained for the correlation between SOC and WAI depends on the selected WAI items. The highest correlations were obtained for WAI 7 "Mental resources", which corresponds to physical resources (r from 0.46 to 0.61) and WAI 1 reflecting the current work ability as compared with lifetime best (r from 0.36 to 0.52). The correlations between the SOC and WAI 3 (standing for the number of diseases diagnosed by a physician) are either statistically insignificant (coronary heart disease, diabetes) or much lower than the earlier presented values. The correlations between SOC and WAI 5 (denoting the sick leave during the last 12 months) are also low.

TABLE 10.1
WAI in Different Groups of Persons with Chronic Diseases

		Statistics	SD
Hypertension	**Mean**	**37.04**	5.99
	Minimum	19.00	
	Maximum	47.00	
Coronary heart diseases	**Mean**	**33.48**	6.04
	Minimum	23.50	
	Maximum	45.00	
Diabetes	**Mean**	**34.68**	6.03
	Minimum	23.00	
	Maximum	46.00	
Osteoarthritis	**Mean**	**35.85**	5.81
	Minimum	19.00	
	Maximum	47.00	

TABLE 10.2
SOC among Workers with Different Chronic Diseases

| | Mean | SD | 95% Confidence Interval for the Mean Value | | Min | Max | Kruskal–Wallis Test for Independent Samples |
			Dolna Granica	Górna Granica			
Hypertension	135.04	20.57	131.96	138.12	68	193	p < 0.000
Coronary heart disease	124.88	18.31	118.49	131.27	91	169	
Diabetes	122.43	20.53	116.34	128.53	68	172	
Osteoarthritis	132.18	20.42	128.55	135.81	70	193	
Total	131.65	20.73	129.56	133.75	68	193	
Healthy persons in a Polish sample (N = 523, Pasikowski, 2001)	132.36	20.39			47	199	

TABLE 10.3
SOC among Workers with Different Chronic Diseases by Gender (F – Female, M – Men)

	Mean	SD	F (ANOVA)	p
		Hypertension		
F	131.68	19.57	4.826	**0.029**
M	138.47	21.02		
		Coronary Heart Disease		
F	120.20	14.70	ns*	ns
M	125.69	18.97		
		Diabetes		
F	120.92	18.52	ns	ns
M	124.09	22.85		
		Osteoarthritis		
F	129.18	18.96	9.316	**0.003**
M	141.71	22.01		

*not statistically significant.
Bold means statistically significant value/difference, p < 0.05.

TABLE 10.4

SOC among Workers with Different Chronic Diseases by Age

	Mean	SD	F (Anova)	p
		Hypertension		
Up to 55	138.66	18.66	3.956	**0.048**
More than 55	132.43	21.56		
		Coronary Heart Disease		
Up to 55	120.20	14.70	ns	ns
More than 55	125.69	18.97		
		Diabetes		
Up to 55	128.61	17.12	ns	ns
More than 55	118.46	21.83		
		Osteoarthritis		
Up to 55	134.93	18.96	ns	ns
More than 55	129.91	21.42		

Bold means statistically significant value/difference, $p < 0.05$.

TABLE 10.5

Correlation between SOC and Work Ability (WAI) in Persons with Different Chronic Diseases (Pearson's Correlation Coefficient)

	WAI 1	WAI 2	WAI 3	WAI 4	WAI 5	WAI 6	WAI 7	WAI
SOC – hypertension	**0.47****	**0.43****	**0.20****	**0.17***	**0.21****	**0.35****	**0.51****	**0.47****
SOC – coronary heart disease	**0.42***	0.26	0.02	0.11	0.06	0.22	**0.61****	0.35^
SOC – diabetes	**0.52****	**0.42****	–0.02	0.21	0.28	**0.47****	**0.54****	**0.57****
SOC – osteoarthritis	**0.36****	**0.38****	**0.26****	**0.20***	0.12	**0.26****	**0.46****	**0.42****

**$p < 0.01$,
*$p < 0.05$,
^$p < 0.10$.
Bold means statistically significant value/difference, $p < 0.05$.

DISCUSSION AND PRACTICAL IMPLICATIONS

As confirmed by the literature review, attempts have already been made to use WAI for work ability assessment in persons with chronic diseases. So far, WAI was used to assess work ability in people with: rheumatoid arthritis (Łastowiecka et al. 2006; Jędryka-Góral et al. 2006; de Croon et al. 2005), cardiovascular disorders in total: hypertension and coronary heart disease in papers by Eskelinen et al. (1991) and Jędryka-Góral et al. (Jędryka – Góral et al. 2006), systemic scleroderma (Sandqvist et al. 2010), chronic nonspecific pain syndromes (Brox et al. 2005; Neupane et al. 2011; de Vries et al. 2012; Bugajska and Sagan 2014), cancer (Bottcher et al. 2013), depression (Boschman et al. 2014), dermatomyositis (Regardt et al. 2015), allergic skin conditions (Plat et al. 2012) and vibration syndrome (Edlund et al. 2012). WAI was also used as the indicator of treatment efficacy (Hoving et al. 2009) or to compare the efficacy of various approaches to rehabilitation (Księżopolska-Orłowska et al. 2016). The above-mentioned papers report a strong negative correlation between the prevalence of chronic diseases and WAI. Given this fact, the study outcome reported by Plat et al. (2012) seems intriguing, as the obtained data don't indicate decreasing values of work ability in firemen presenting with chronic diseases. The authors, criticizing these results, explain that the chronic diseases reported by firemen include dermal conditions with symptoms which are not likely to impair psychophysical skills, necessary to perform their job (Plat et al. 2012).

Based on the literature review, we can also conclude that chronic diseases evidently reduce work ability measured using WAI. The values obtained in this project additionally indicate that work ability levels are differentiated, depending on the type of chronic diseases. In our study, the participants with coronary heart disease and osteoarthritis were characterized by lower work ability levels.

The results also indicated that the SOC in the studied sample was differentiated according to chronic diseases. Since the authors of the questionnaire measuring the SOC consciously ignored the reference values (norms), it is advised to read the reports on other studies using this tool to obtain information on relative levels of the SOC in the studied groups. The analyses conducted by Pasikowski (2001) in samples of healthy people (overall N = 523) allow such comparison. It turns out that the mean values obtained in the groups of workers with arterial hypertension and osteoarthritis (135.04 and 132.18) are similar and almost identical as the values obtained in healthy individuals (132.36). The values obtained from the groups with coronary heart disease and diabetes were lower than those obtained from healthy participants. The average result of this study partly confirms the findings that chronic diseases are associated with decreased SOC levels (Eriksson and Lindström 2006). There are also studies performed on large populations (Eriksson et al. 2013) whose results don't confirm the above finding. If the disease advancement level was confirmed, it would be easier to explain this phenomenon.

The study results show a strong correlation between work ability in working people with chronic diseases with their individual characteristic, which is the SOC. The correlation is robust and confirmed by the value obtained for r Pearson (0.48). The value referring to the correlation depended on the selected WAI item. Higher correlation coefficients were found for WAI items corresponding to well-being, especially mental resources, whereas lower values are related to objective indicators of health status. This result is in agreement with the earlier findings, indicating that the SOC mainly depends on mental health and less – on physical health (Eriksson and Lindström 2006). Here, it is worth noting that the strongest correlations were found between SOC and WAI – assessment of mental resources (it is of note that some questions in SOC questionnaire are similar to those corresponding to mental resource assessment in WAI questionnaire). We may thus conclude that the SOC is

mainly related to subjective measures of mental comfort and weakly or not correlated at all with objective health indicators. The relationship between SOC and lower
levels of emotional tension are also confirmed by the results obtained in a sample of
workers from Japan (Urakawa et al. 2012).

As mentioned at the beginning, the research conducted so far has not considered
the direct relationships between the two reported phenomena, which are work ability in persons with chronic illnesses, on the one hand, and an individual trait in the
form of a SOC, on the other hand. The strong correlation between such phenomena,
disclosed in this study, suggests that some preventive measures should be implemented to improve work in ability workers suffering from chronic diseases. Certain
therapeutic activities, enhancing the SOC, such as comprehensibility, manageability
and sense may prove effective. The current long-term interest in SOC from researchers has prompted them to develop a substantial number of interventions aimed at
enhancement of this characteristic. These include such undertakings as educational
programs, facilitating comprehension and disease-coping (Odajima et al. 2017),
interventions focused on the increase of activity levels and coping with difficult situations (Tan et al. 2016), mindfulness training (Ando et al. 2011; Weissbecker et al.
2002), or various forms of psychotherapy (Mayer and Boness 2011). All these actions
lead to enhancement of the SOC. However, there have been no studies on the favourable effect of these actions on work ability. Nevertheless, given the confirmed relationship between the two studied phenomena (SOC and work ability), we may expect
a positive outcome also in this area.

REFERENCES

Agardh, E. E., A. Ahlbom, T. Andersson et al. 2003. Work stress and low sense of coher-
     ence is associated with type 2 diabetes in middle-aged Swedish women. *Diabetes Care*
     26(3):719–724.
Ando, M., T. Natsume, H. Kukihara, H. Shibata, and S. Ito. 2011. Efficacy of mindfulness-
     based meditation therapy on the sense of coherence and mental health of nurses. *Health*
     3(2):108–122. DOI: 10.4236/health.2011.32022.
Antonovsky, A. 1987. *Unraveling the mystery of health: how people manage stress and stay
     well.* San Francisco: Jossey-Bass.
Benyamini, Y., I. Roziner, U. Goldbourt, Y. Drory, Y. Gerber, and Israel Study Group on
     First Acute Myocardial Infarction. 2013. Depression and anxiety following myocardial
     infarction and their inverse associations with future health behaviors and quality of life.
     Ann Behav Med 46(3):310–321.
Boschman, J. S., H. F. van der Molen, M. H. W. Frings-Dresen, and J. K. Sluiter. 2014.
     The impact of common mental disorders on work ability in mentally and physically
     demanding construction work. *Int Arch Occup Environ Health* 87(1):51–59.
Bottcher, H. M., M. Steimann, A. Ullrich et al. 2013. Work-related predictors of not returning
     to work after inpatient rehabilitation in cancer patients. *Acta Oncol* 52(6):1067–1075.
Brox, J. I., K. Storheim, I. Holm, A. Friis, and O. Reikersa. 2005. Disability, pain, psychologi-
     cal factors and physical performance in healthy controls, patients with sub-acute and
     chronic low back pain: a case-control study. *J Rehabil Med* 37(2):95–99.
Bugajska, J., and A. Sagan. 2014. Chronic musculoskeletal disorders as risk factors for reduced
     work ability in younger and ageing workers. *Int J Occup Saf Ergon* 20(4):607–615.

Chumbler, N. R., K. Kroenke, S. Outcalt et al. 2013. Association between sense of coherence and health-related quality of life among primary care patients with chronic musculoskeletal pain. *Health Qual Life Outcomes* 11(1):126. Article Number 2016.

de Croon, E. M., J. K. Sluiter, T. F. Nijssen et al. 2005. Work ability of Dutch employees with rheumatoid arthritis. *Scand J Rheumatol* 34(4):277–283.

de Vries, H. J., M. F. Reneman, J. W. Groothoff, J. H. Geertzen, and S. Brouwer. 2012. Workers who stay at work despite chronic nonspecific musculoskeletal pain: do they differ from workers with sick leave? *J Occup Rehabil* 22(4):489–502.

Edlund, M., L. Gerhardsson, and M. Hagberg. 2012. Psychical capacity and psychological mood in association with self-reported work ability in vibration-exposed patients with hand symptoms. *J Occup Med Toxicol* 7(1):22. Article Number 22.

Eriksson, A. K., M. Van den Donk, A. Hilding, and C. G. Östenson. 2013. Work stress, sense of coherence, and risk of type 2 diabetes in a prospective study of middle-aged Swedish men and women. *Diabete Care* 36(9):2683–2689.

Eriksson, M., and B. Lindström. 2006. Antonovsky's sense of coherence scale and the relation with health: a systematic review. *J Epidemiol Community Health* 60(5):376–381.

Eriksson, M., and B. Lindström. 2007. Antonovsky's sense of coherence scale and the relation with quality of life: a systematic review. *J Epidemiol Community Health* 61(11):938–944.

Eskelinen, L., A. Kohvakka, T. Merisalo, H. Heikki, and G. Wagar. 1991. Relationship between the self-assessment and clinical assessment of health status and work ability. *Scan J Work Environ Health* 17(suppl 1):40–47.

Hoving, J. L., G. M. Bartelds, J. K. Sluiter et al. 2009. Perceived work ability, quality of life and fatigue in patients with rheumatoid arthritis after a 6-month course of TNF inhibitors: prospective intervention study and partial economic evaluation. *Scand J Rheumatol* 38(4):246–250. DOI: 10.1080/03009740902748264.

Jędryka-Góral, A., J. Bugajska, E. Łastowiecka et al. 2006. Work ability in ageing workers suffering from chronic diseases. *Int J Occup Saf Ergon* 12(1):17–30.

Jueng, R. N., D. C. Tsai, and I. J. Chen. 2016. Sense of coherence among older adult residents of long-term care facilities in Taiwan: a cross-sectional analysis. *PloS One* Jan 11;11(1):e0146912.

Księżopolska-Orłowska, K., A. Pacholec, A. Jędryka-Góral et al. 2016. Complex rehabilitation and the clinical conditions of working rheumatic arthritis patients – does cryotherapy always overtop traditional rehabilitation? *Disabil Rehabil* 38(11):1034–1040.

Łastowiecka, E., J. Bugajska, A. Najmiec et al. 2006. Occupational work and quality of life in osteorthritis patients. *Rheumatol Int* 27(2):131–139.

Mayer, C. H., and C. Boness. 2011. Interventions to promoting sense of coherence and transcultural competences in educational contexts. *Int Rev Psych* 23(6):516–524.

Neupane, S., P. Virtanen, P. Leino-Arjas et al. 2013. Multi-site pain and working conditions as predictors of work ability in a 4-year follow-up among food industry employees. *Eur J Pain* 17(3):444–451.

Odajima, Y., M. Kawaharada, and N. Wada. 2017. Development and validation of an educational program to enhance sense of coherence in patients with diabetes mellitus type 2. *Nagoya J Med Sci* 79(3):363–374. DOI:10.18999/nagjms.79.3.363.

Pasikowski, T. 2001. Kwestionariusz poczucia koherencji dla dorosłych (SOC-29). In: *Health – stress – resources*, ed. Sęk, H., T. Pasikowski, 71–85. Poznań: Wydawnictwo Fundacji Humaniora.

Plat, M. C., M. H. W. Frings-Dresen, and J. Sluiter. 2012. Impact of chronic disease on work ability in ageing firefighters. *J Occup Health* 54(2):158–163.

Regardt, M., E. Welin Henriksson, J. Sandqvist, I. E. Lundberg and M. L. Schult. 2015. Work ability in patients with polymyositis and dermatomyositis: an explorative and descriptive study. *Work* 53(2):265–277.

Sandqvist, G., A. Scheja, and R. Hesselstrand. 2010. Pain, fatigue and hand function closely correlated to work ability and employment status in systemic sclerosis. *Rheumatology* 49(9):1739–1746.

Silarova, B., I. Nagyoval, J. Rosenberger et al. 2012. Sense of coherence as an independent predictor of health-related quality of life among coronary heart disease patients. *Qual Life Res* 21(10):1863–1871.

Tan, K. K., S. W. Chan, W. Wang, and K. Vehviläinen-Julkunen. 2016. A salutogenic program to enhance sense of coherence and quality of life for older people in the community: a feasibility randomized controlled trial and process evaluation. *Patient Educ Couns* 99(1):108–116. DOI: 10.1016/j.pec.2015.08.003. Epub 2015 Aug 11.

Tuomi, K., J. Ilmarinen, A. Jahkole, L. Karajaninne, and A. Tulkki. 1998. *Work Ability Index.* Helsinki: Finnish Institute of Occupational Health.

Urakawa, K., K. Yokoyama, and H. Itoh. 2012. Sense of coherence is associated with reduced psychological responses to job stressors among Japanese factory workers. *BMC Res Notes* 2012(5):247.

Vooij, M., M. C. J. Leensen, J. L. Hoving, H. Wind, and M. H. W. Frings-Dresen. 2018. Value of work for employees with a chronic disease. *Occup Med-Oxf* 68(1):26–31.

Weissbecker, I., Salmon, P., J. L. Studts et al. 2002. Mindfulness-based stress reduction and sense of coherence among women with fibromyalgia. *J Clin Psychol Med Settings* 9(4):297–307. DOI: 10.1023/A:1020786917988.

Wrześniewski, K., and D. Włodarczyk. 2012. Sense of coherence as a personality predictor of the quality of life in men and women after myocardial infarction. *Kardiol Pol* 70(2):157–164.

11 Adjustment of Work Organization and Working Conditions to the Needs of Persons with Chronic Diseases – Results of Research

Karolina Pawłowska-Cyprysiak, Katarzyna Hildt-Ciupińska, and Joanna Bugajska
Central Institute for Labour Protection –
National Research Institute

CONTENTS

INTRODUCTION

One quarter of the working population in Europe suffer from chronic diseases, and 8% growth in the prevalence of such health conditions was noted between 2010 and 2017. This increasing trend will be maintained due to the ageing population and the fact that workers aged 50 years and above twice as often present with chronic diseases as compared with workers aged below 35 years (Eurofound 2019). Considering the prevalence of each disease, the most widespread ones include low back pain, arterial hypertension, pain in the neck or the middle back pain, osteoarthritis, migraine, coronary heart disease, allergy, thyroid diseases or diabetes. In the population aged 50 years and above, the percentage of people with arterial hypertension and back or neck pain is still increasing. People aged 60 years and above in turn, apart from cardiovascular system disorders (arterial hypertension, coronary heart disease) and problems with joints, also develop type II diabetes or urinary tract diseases.

This population is also at risk of depression and cerebrovascular stroke (GUS 2016). Given the fact that the prevalence of chronic diseases increases with age and, at the same time, the working population grows older, we can assume that workers will increasingly often present with chronic diseases (Koolhaas et al. 2014). Therefore, it is essential to know the needs of this group of workers regarding work organization and environmental conditions, in order to employ adequate measures to maintain their work ability.

People with Chronic Diseases in the Labor Market

The duration of a chronic disease has a marked effect on labor force participation rate and people's desire to be occupationally active. This correlation is much stronger in the male population as compared to women (Rijken et al. 2013). People with chronic diseases are less often occupationally active than healthy people. However, if they work regardless their health problems, they are more frequently on long-term sick leave (Boot et al. 2013).

The factor responsible for the high rate of absenteeism in workers with chronic diseases is self-perception of their health. The results presented by Boot indicate that self-rating of health status was correlated with a higher frequency of absenteeism due to diseases (Boot et al. 2011). The strength of the above correlations, however, decreases after adding work limitations to the model. The authors thus indicate that both self-rating of health and work limitations are the factors explaining absenteeism at a similar level. Moreover, the presence of a chronic disease is associated with lower work ability rating and minimizes the chances for employment (Koolhaas et al. 2014). The results obtained by de Boer et al. (2018) also indicate that chronic diseases are predictors of earlier resignation from job.

Splitting headaches, diabetes and respiratory and digestive system disorders as well as mental disorders are connected with a higher risk of qualifying patients for disability benefits. This effect is modified by psychosocial working conditions. Social support and more autonomy at work together with lower psychological demands decrease the risk of being qualified for disability benefits due to chronic conditions (Leijten et al. 2015).

In the case of patients with chronic diseases who remain employed, work is perceived as a great value having a therapeutic and motivating effect, and as a source of income. For such persons, having a job is very important. They are motivated to remain employed as long as possible and are eager to take actions to meet this goal (de Vries et al. 2011).

Problems Encountered by People with Chronic Diseases at Work

People with chronic diseases may experience many everyday life problems including those connected with job performance. These problems involve physical, cognitive or sensory limitations; difficulties in meeting job requirements; fatigue; and other problems related to their illness. The disease may also affect their mental sphere and result in depression, feeling ashamed and having difficulties in communication with other people; such problems may be aggravated due to the lack of the manager's and coworker's support (Varekamp et al. 2013).

Job requirements and working time may be too burdensome for workers with chronic diseases. In such cases, a worker may be required to change a job title or has no chances for promotion, which, in turn, may contribute to a decreased quality his/her professional life (de Jong et al. 2015). Such circumstances may be due to difficulties in communicating the employer, improper social relations at work or no opportunities for adjustment of working conditions to the worker's health problems (Dekkers-Sánchez et al. 2010).

The actions aimed at finding appropriate solutions for problems with functioning in a workplace and reduction of sick leave days in workers suffering from chronic diseases should be focused on the following: (1) a worker him/herself, (2) workplace and (3) health care (Varekamp et al. 2013). Adjustment of the workplace/workstation to one's needs resulting from disease-related functional limitations is important for this group of workers (Boot et al. 2013). It can equalize their chances for getting a job (Koolhaas et al. 2014; Hjärtström et al. 2018).

The aim of the study presented in this chapter was to identify the needs for adjustment of work organization and working conditions to people suffering from chronic diseases, considering a social factor, i.e. support from the employers and coworkers, as well as physical environmental factors in the workplace that may favorably affect and help maintain work ability and involvement in occupational activities in this population.

METHODS

The study was carried out using a qualitative approach, involving the application of structured individual extended surveys. The surveys were based on the tailored questionnaire. It consisted of general questions (the respondent's age, gender and job), 9 questions related to health and 13 questions related to the respondent's job.

The first part of the questionnaire, pertaining to health status assessment, included the following items:

- Assessment of current health status
- The respondent's chronic diseases and their duration
- Other diseases
- The type of applied treatment and observing physician's recommendations
- Problems associated with current health status and the impact of these problems on functioning at work.

The part related to the situation at work included the following items:

- The number of weekly working hours
- Satisfaction with working conditions
- Assessment of the adjustment of working conditions to the respondent's health problems
- Identifying the factors in a workplace which should be limited and eliminated

- Identifying the issues that depend on the respondent during work performance
- Stating whether the employer and coworkers should be informed about the respondent's comorbidities
- Stating how the employer and coworkers could help the respondents preform his/her duties
- Stating whether the respondent's health status affects his/her work performance
- Self-assessment of work ability in 2 years' perspective, considering the respondent's health status.

During the direct interviews, the Work Ability Index (WAI) questionnaire was not applied; however, the question concerning the 2 years' perspective for remaining employed is inspired by WAI, including the question concerning their own prognosis of work ability for the next 2 years. Participation in the study was voluntary. The interview was carried out "face-to-face" in the place and time suitable for the respondents. The duration of each interview ranged from 30 to 60 minutes.

MATERIAL

Selection of the sample was deliberate and based on having of one of the four selected diseases, namely (1) coronary heart disease, (2) arterial hypertension, (3) osteoarthritis (affecting spinal, knee, elbow and carpal joints) and (4) type II diabetes. The sample comprised 60 persons, divided into 4 equipotent groups of 15 participants.

RESULTS

SAMPLE CHARACTERISTICS

Considering the sample characteristics, it is of note that the youngest groups of participants comprised people with osteoarthritis, whereas the oldest group comprised people with coronary heart disease. The participants with the longest disease duration included those with arterial hypertension, whereas in the remaining group, the average disease duration was similar. The number of weekly working hours was similar in each group. Table 11.1 presents characteristics of each group participating in the survey, by age, gender, average disease duration and the number of weekly working hours.

ASSESSMENT OF HEALTH STATUS

The highest self-rating of health status was noted in workers with osteoarthritis; 15 workers rated their health as *very good, good* and *satisfactory*. A similar rating was reported by the workers with arterial hypertension. *Poor* health was reported by only one participant. The workers with diabetes and coronary heart disease obtained lower ratings of their health status. In both groups, 100% of workers rated their health as *good* and *satisfactory*. None of the respondents with diabetes and coronary heart disease rated their health as very good and none of them rated their health as *very poor* or *poor*. (Table 11.2).

TABLE 11.1
Characteristics of the Studied Sample (N=60)

	Age	Gender	Average Disease Duration	Number of Weekly Working Hours
Diabetes	57.7	8 women 7 men	6.5	39.3
Coronary heart disease	63.2	6 women 9 men	6.6	40.7
Arterial hypertension	54.8	8 women 7 men	12.1	40–45
Osteoarthritis	41.8	9 women 6 men	7.5	40

TABLE 11.2
Comparison of Subjective Ratings of Health Status in Workers with Chronic Diseases (N=60)

	Very Good	Good	Satisfactory	Poor	Very Poor
Diabetes	0	8	7	0	0
Coronary heart disease	0	7	8	0	0
Arterial hypertension	1	5	8	1	0
Osteoarthritis	2	7	6	0	0

SITUATION AT WORK AND SATISFACTION WITH WORKING CONDITIONS IN THE STUDIED SAMPLE

Considering the work content in all four groups of workers surveyed using direct interviews, the vast majority of the sample were people performing mental work. The jobs of each group of workers are presented as follows:

- Persons with diabetes – physician (2), sales representative (1), economic technician (1), secretary (2), psychologist (1), pharmacist (2), lawyer (1), specialist (1), salesman (2), car mechanic (1) and scientist (1).
- Persons with coronary heart disease – laboratory technician (2), pharmaceutical technician (2), teacher (1), doctor (3), computer science (1), seller (3) and engineer (3).
- Persons with hypertension – researcher (2), car mechanic (1), doctor (1), lawyer (1), secretary (1), HR manager (1), salesperson (3), driver (2) and specialist (3).
- Persons with osteoarthritis – computer graphic designer (1), engineer (4), accountant (2), lawyer (1), psychologist (1), administrative worker (4) and nurse (1).

TABLE 11.3
Comparison of Subjective Assessment of Working Conditions in Workers with Chronic Diseases (N=60)

	Yes, I'm Very Satisfied	Yes, I'm Satisfied	I'm not very Satisfied	I'm not Satisfied at All	It's Difficult to Say
Diabetes	2	11	2	0	0
Coronary heart disease	6	9	0	0	0
Arterial hypertension	2	10	3	0	0
Osteoarthritis	1	9	5	0	0

The majority of the surveyed workers were satisfied with their working conditions, regardless of their diseases. Most of the positive answers, *yes, I am satisfied* and *yes, I am very satisfied with my working conditions*, were noted in the group of workers with coronary heart diseases including 5 positive ratings among 15 respondents; next, in the group with diabetes, including 13 positive ratings among 15 respondents; and next, in the group with arterial hypertension, including 12 positive ratings among 15 respondents. The lowest number of positive ratings was noted in workers with osteoarthritis, including 10 positive ratings among 15 respondents.

It is of note that the majority of respondents (n = 5) who were *not very satisfied* with their working conditions suffered from osteoarthritis (Table 11.3).

Subjective Assessment of the Possibility to Stay at a Current Job within the Next 2 Years from the Point of View of the Worker's Medical Condition

Most of the respondents (N = 58) asked about the prognosis for their employment within the next 2 years reported that they would be able to work in their profession within the next 2 years. All the workers with diabetes (N = 15), arterial hypertension (N = 15) and osteoarthritis (N = 15) reported that they would remain employed until then. In the group with coronary heart disease, 13 respondents reported that they might remain employed within the next 2 years, whereas 2 respondents were not sure they would.

The Factors That Should Be Limited or Eliminated from the Working Environment According to Respondents

In workers with **diabetes**, the most often mentioned factors in the working environment, which should be eliminated and/or reduced due to their disease, included prolonged sitting and painful/tiring body position while working; according to workers with **coronary heart disease,** such factors included prolonged sitting, prolonged standing and assuming a painful/tiring body position; many respondents with **arterial hypertension** reported such factors as prolonged sitting, too fast work pace, prolonged standing and assuming a painful/tiring body position while working, an excessively fast work pace, too high ambient temperature in the workplace, inadequate natural lighting, stressful contacts with coworkers, bending, assuming a painful/tiring body position while working and repeatable hand or arm movements; according to the respondents with **osteoarthritis**, such factors included assuming a painful/tiring body position while working, prolonged sitting, prolonged standing, repeated hand or arm movements, lifting or moving heavy loads, working with a computer, too low or too high ambient temperatures (in the workplace/workstation or outside), noise and stressful contacts with customers/clients.

ADJUSTMENT OF THE WORKSTATION TO WORKERS' NEEDS RESULTING FROM THEIR MEDICAL CONDITIONS

The vast majority of respondents reported that their working conditions were adjusted to their needs resulting from their health problems, but they also suggested solutions allowing even more satisfactory improvement of their comfort at work.

The results indicate that

- The workers with diabetes pointed to the need for more frequent and longer breaks at work, and for eating warm meals during working hours. ("I have to eat regularly every three hours, and I would like to have such an opportunity", "I would like to be able to eat something whenever Ixneed".)
- The workers with coronary heart disease suggested the need for the reduction of work-related stress. ("I would like to undergo stress-management training", "I would like to have less stressful duties".)
- The workers suffering from arterial hypertension suggested the need for reduction of work-related stress ("I wish my work would be more organized and less stressful", "Less stressful situations resulting from regular contacts with other people", "Less tension, less stress") and improvement of physical conditions ("I would also like to have better ventilation in the room and more natural light", "I would like to have a more functional chair").
- The workers suffering from osteoarthritis suggested the need for workstation ergonomics improvement ("A desire to have a desk with an adjustable countertop, a desire to have a foot rest because the leg should be straightened", "It is not recommended for a person with discopathy to sit all the time at work. The spine is loaded. Besides, carrying a backpack with a laptop has a negative impact on my spine", "I could have less stress", "Reduction of the amount of work requiring long hours of sitting in one position.", "Breaks for gymnastics should be longer", "Work with a computer can be done at home (place of living)".

POSSIBLE IMPACT ON WORK ORGANIZATION

Workers with diabetes reported their possible impact on such organizational factors in their working environments as the order of individual work-related activity performance, the manner of work performance and, to a lower extent, the pace of work. The respondents with **coronary heart disease** reported their possible impact on the manner of work performance, the length and frequency of breaks at work, and the sequence of occupational duty performance. The respondents with **arterial hypertension** reported their possible impact on the sequence of performing work-related activities, the manner and pace of work performance and the length of rest breaks. According to workers with osteoarthritis, the organizational factors in the working environment that may depend on workers, the manner of work performance, work pace and the frequency and length of rest breaks at work.

SOCIAL WORKING ENVIRONMENT AND HEALTH-RELATED NEEDS

Providing Information on Health Status

The analysis of the results has shown that most of the respondents with diabetes, coronary heart disease and osteoarthritis (10, 10 and 12 workers, respectively) believe that both the employer and workers should be informed/about their illnesses. In the group of workers with arterial hypertension, only seven respondents expressed this opinion.

The following answers were given to the question why such information should be disclosed.

The respondents with arterial hypertension believe that "the workmates should know so as to react in such cases and help if something happens. Also, when it is necessary to call an ambulance, they will be able to inform the ambulance service personnel about the patient's condition", "If medical assistance from third parties is needed, the information about the illnesses and medication taken by the patient is very important". The respondents who did not report such a need reported no influence of a given medical condition on their work performance:

> "The disease doesn't require informing the coworkers. Nor does it affect my job performance." This disease has no impact on work performance, "It's no need", "Because it shouldn't affect work performance".

According to the respondents with osteoarthritis, the information about the medical condition should be disclosed because "The employer could make suitable improvements in the working environment, e.g. adjust the desk or chair", "They are always ready to help in difficult situations, the supervisor may reorganize work", "Knowing about my illness, the supervisor could better adjust workload to my health problem", "If needed, helping carrying heavy files with documents and several-minute rest breaks if pain intensifies". The respondents who reported no need for help justified their choice by no effect of the disease on the interpersonal relations at work or job itself, or the lack of confidence in positive results of information disclosure: "The disease has no impact on work and interpersonal relations, therefore it's useless to talk about it" "Because it won't change a lot" and "There's no such a need". One of the respondents believed that it is an individual concern, depending on the work content "It also depends on the kind of work – if someone carries loads every day, he/she should inform the employer about it".

According to the workers suffering from diabetes, it is important to inform the employer/coworkers about their condition as it is likely to affect this worker's safety at work. ("The workers know about my disease and how to help me in case of drop in blood sugar level", "I feel safe because I know that other people know".) The persons who reported no need to inform the supervisor or colleagues about their medical conditions most often justified it by fear of being fired or thinking that the disease is his/her problem. ("They might think that I'm not fit for this job", "They don't under-stand my illness", "One can live with it", "My illness is my problem", "I am afraid that I will be fired when I tell them".)

The respondents suffering from coronary heart disease most often reported the need of concern about workers' safety being the most frequent reason of informing the supervisor about the worker's medical condition. ("The fact that my colleagues know about my condition makes me feel safe", "My colleagues know how to react when something happens to me because they know about my conditions", "Other people know how to help me when I need help".) The workers in this group who reported no need of informing the supervisor or the coworkers about their conditions reported that it is their private matter; therefore, there is no need to inform anyone about it. ("I don't want anyone to know about my complaints", "I don't want me and my condition to be a subject of any gossip".)

Help with Daily Activities

The respondents were also asked whether it was possible to get help in daily activities from their employer or coworkers. Most of the respondents suffering from diabetes, coronary heart disease, arterial hypertension and osteoarthritis reported no need or possibility to do so. They also reported no effect of their health or stability of their medical condition on their quality of work and their occupational efficiency.

The Effect of Work on Health According to the Respondents

Most of the respondents suffering from diabetes (n = 11) were not able to determine how occupational activities affected their health. Among the remaining respondents from this group, two reported a positive effect of work on health, whereas two other respondents reported no impact of work on their health. ("My condition does not affect my work", "I've been ill for years and I don't know how to cope with my condition".)

Most of the respondents suffering from **coronary heart disease** (n = 11) reported no effect of work on their health or were not able to determine this effect. The remaining four respondents reported a favorable effect of their occupational duties on their health (motivation for living and involvement in the activities of daily life). (e.g. "I work and I'm not motivated for other activities", "The fact that I work makes me believe that I will manage to do a lot of things".)

Among the respondents suffering from **arterial hypertension,** the vast majority (n = 13) reported an adverse effect of work on their health. They gave the following reasons for such answers: "Sometimes, in stressful situations, it takes me a long time to think about them at home, wandering what errors I could have committed at work being under stress and in a hurry".

Two respondents reported a favorable effect of work on their health because they enjoy working due to financial and social reasons.

A vast majority of respondents suffering from **osteoarthritis** (n = 13) reported an adverse effect of work on their health. They justified their answers in the following way: "Sometimes I have too much work to do and I'm stressed", "Noise doesn't allow me to concentrate", "High stress combined with excessive mental load adversely affects my health". Two respondents reported a favorable effect of work on their health, whereas two other respondents reported that work sometimes favorably and other times adversely affected their health. None of them, however, justified their answers.

DISCUSSION

The results of the interviews indicate that the majority of respondents were generally satisfied with their present working conditions regardless their health-related problems. Most of the answers were either "satisfied" or "very satisfied". Most of the "satisfied" and "very satisfied" respondents were noted in the group of workers with coronary heart disease with 100% of overall positive ratings and in workers suffering from diabetes with 86.6% of overall positive ratings.

As it has already been mentioned, prevention of the problems with functioning at work, reduction of disease-related absenteeism in workers suffering from chronic diseases and sustaining the work ability of these workers may be developed in three main directions: (1) focusing on the worker only, (2) focusing on the workplace and (3) focusing on health care. The description and discussion of the results obtained from the extended surveys will be focused on two of the three directions of actions, i.e. those focused on the worker only and those focused on the workplace.

ACTIONS FOCUSED ONLY ON THE WORKER

The responses obtained from this survey indicate what actions are expected by the studied workers from their employer. Respondents reported a desire for more autonomy at work, possibility to decide about the pace of and manner of work performance and pointed to the need for limiting the actions and circumstances that may contribute to or intensify stress, such as working in exposure to mental load and excessive job requirements. Workers suffering from chronic diseases, participating in our research, pointed to the need for workplace adjustment to their health problems, working time reduction, a possibility to perform work at home and to adjust their work pace to their needs, the workers' impact on work and a possibility to decide on it. These directions of actions are similar to those reported by the participants of Varekamp's survey (Varekamp and van Dijk 2010).

One of the participants of the survey pointed out that adverse circumstances encountered in the workplace and the committed errors influence the workers' thoughts after working hours. Such problems are still present outside the workplace and result in imbalance between work and private life. On the other hand, however, the answers indicating that the respondent's work has no adverse impact on their health and, on the contrary, it favorably affects their health are of note. They more often referred to physical than to mental health; this issue, however, is very interesting. For the respondents, work is the driving force and motivation for acting in other spheres of life. Their positive attitudes towards work resulted in motivation for undertaking other extracurricular activities.

Given the above, we should pay attention to the two trends related to the effect of work on private life. On one hand, these include the adverse effect of one sphere of life on another and the resulting stress ("Sometimes, when I face stressful situations, I think for a long time about them and about the errors I could have committed working in stress and under pressure"). On the other hand, however, the respondents report that their private life is enriched by their work ("I work and thus I'm motivated for other activities"). Nevertheless, in both above-mentioned cases a wider research is recommended to find out whether such phenomena are regularly encountered in larger groups of people or whether they are person-specific and depend on personal characteristics, current health status (disease), work-related circumstances, etc. Furthermore, determining the factors affecting such different perceptions of the relationship between work and private life by a group of persons suffering from chronic diseases seems interesting. In most cases, the respondents reported they were convinced about the need to inform the supervisors and coworkers about their health status, using the argument that safety and work comfort should be the priority for workers with chronic diseases and their coworkers. They do not expect help or replacement in work performance.

According to Butler and Modaff (2016), the reasons for informing the supervisors and coworkers about health status by workers suffering from chronic diseases include the need to justify their absenteeism or health problems, or to facilitate their continuation of employment. Providing information about each disease is important for a proper work organization, and in cases of life-threatening situations resulting from, e.g., hypoglycemia or sudden increase in blood pressure (sudden high blood pressure), quick intervention is essential. In people with chronic diseases, the reasons for not informing the supervisors and/or coworkers about their health problems may be due to the unfriendly atmosphere in their workplace, stereotypes towards and stigmatization of such persons in and outside of the workplace. In consequence, it may lead to poor social awareness of such health problems and unrealistic employers' expectations towards workers with chronic diseases (de Jong et al. 2015).

WORKPLACE-FOCUSED ACTIVITIES

According to the majority of respondents, their workplace is not adjusted to their health-related needs. Some persons, however, mentioned their expectations concerning physical adjustment of the workplace aimed at increasing their work comfort. Such expectations included improvements of working conditions by reducing ambient temperature and providing ventilation and, thus, increasing the thermal comfort, providing greater access to natural lighting at the workstation, and ergonomic solutions in the workplace, e.g. height-adjustable desks and footrest support.

Typically, the workers suffering from osteoarthritis and arterial hypertension, as compared with those suffering from coronary heart disease and diabetes, named more factors to be eliminated from the working environment, being excessive burdens placed on them. On the one hand, it may indicate that the working environment reported by the respondents with coronary heart disease, participating in the study, was better adjusted to their psychophysical potential. On the other hand, however, one may assume that from the point of view of physical workstation adjustment involving the application of ergonomic solutions, there are more options for workers with osteoarthritis. In this group of workers, ergonomic solutions have a marked effect on work comfort and do not exacerbate their symptoms.

The literature review reported by Hoving et al. (2013) has shown that employers have numerous opportunities for adjustment of the working conditions and working environment to the needs of workers with chronic diseases. These include ergonomic solutions, flexible working hours, transport to and from the workplace, education and training or assistance in career planning, acceptance, support and understanding, cooperation between the employer and workers with chronic diseases involving implementation of and discussion about available solutions or providing opportunities for making decisions on working time and approaches to work performance.

REFERENCES

Boot, C. R. L., L. L. J. Koppes, S. N. K. van den Bossche, J. R. Anema, and A. J. van der Beek. 2011. Relation between perceived health and sick leave in employees with a chronic illness. *J Occup Rehabil* 21(2):211–219.

Boot, C. R. L., S. G. van den Heuvel, U. Bültmann, A. G. E. M. de Boer, L. L. J. Koppes, and A. J. van der Beek. 2013. Work adjustments in a representative sample of employees with a chronic disease in the Netherlands. *J Occup Rehabil* 23(2):200–208. DOI: 10.1007/s10926-013-9444yy.

Butler, J. A., and D. P. Modaff. 2016. Motivations to disclose chronic illness in the workplace. *Qual Res Rep Commun* 17(1):77–84. DOI: 10.1080/17459435.2016.1143387.

de Boer, A. G. E. M., G. A. Geuskens, U. Bültmann et al. 2018. Employment status transitions in employees with and without chronic disease in the Netherlands. *Int J Public Health* 63(6):713–722. DOI: 10.1007/s00038-018-1120-8.

de Jong, M., A. G. E. M. de Boer, S. J. Tamminga, and M. H. W. Frings-Dresen. 2015. Quality of working life issues of employees with a chronic physical disease: a systematic review. *J Occup Rehabil* 25:182–196. DOI: 10.1007/s10926-014-9517-6.

de Vries, H. J., S. Brouwer, J. W. Groothoff, J. H. B. Geertzen, and M. F. Reneman. 2011. Staying at work with chronic nonspecific musculoskeletal pain: a qualitative study of workers' experiences. *BMC Musculoskelet Disord* 12:e126. DOI: 10.1186/1471-2474-12-126.

Dekkers-Sánchez, P. M., H., Wind, J. K., Sluiter, and M. H. W. Frings-Dresen. 2010. A qualitative study of perpetuating factors for long-term sick leave and promoting factors for return to work: chronic work disabled patients in their own words. *J Rehabil Med* 42(6):544–552. DOI: 10.2340/16501977-0544.

Eurofound. 2019. *How to respond to chronic health problems in the workplace?* https://www.eurofound.europa.eu/publications/policy-brief/2019/how-to-respond-to-chronic-health-problems-in-the-workplace (accessed January 28, 2020).

GUS [Główny Urząd Statystyczny]. 2016. *Stan zdrowia ludności Polski w 2014 roku.* https://stat.gov.pl/obszary-tematyczne/zdrowie/zdrowie/stan-zdrowia-ludnosci-polski-w-2014-r-,6,6.html (accessed January 28, 2020).

Hjärtström, C., A. Lindahl Norberg, G. Johansson, and T. Bodin. 2018. To work despite chronic health conditions: a qualitative study of workers at the Swedish Public Employment Service. *BMJ Open* 8(4):e019747. DOI: 10.1136/bmjopen-2017-019747.

Hoving, J. L., M. C. B. van Zwieten, M. van der Meer, J. K. Sluiter, and M. H. W. Frings-Dresen. 2013. Work participation and arthritis: a systematic overview of challenges, adaptations and opportunities for interventions. *Rheumatology* 52(7):1254–1264. DOI: 10.1093/rheumatology/ket111.

Koolhaas, W., J. J. L. van der Klink, M. R. de Boer, J. W. Groothoff, and S. Brouwer. 2014. Chronic health conditions and work ability in the ageing workforce: the impact of work conditions, psychosocial factors and perceived health. *Int Arch Occup Environ Health* 87(4):433–443. DOI: 10.1007/s00420-013-0882-9.

Leijten, F. R. M., A. de Wind, S. G. van den Heuvel et al. 2015. The influence of chronic health problems and work-related factors on loss of paid employment among older workers. *J Epidemiol Commun Health* 69(11):1058–1065.

Rijken, M., P. Spreeuwenberg, J. Schippers, and P. P. Groenewegen. 2013. The importance of illness duration, age at diagnosis and the year of diagnosis for labour participation chances of people with chronic illness: results of a nationwide panel-study in the Netherlands. *BMC Public Health* 13:e803. DOI: 10.1186/1471-2458-13-803.

Varekamp, I., and F. J. H. van Dijk. 2010. Workplace problems and solutions for employees with chronic diseases. *Occup Med-Oxf* 60(SI 4):287–293. DOI: 10.1093/occmed/kqq078.

Varekamp, I., F. J. H. van Dijk, and L. E. Kroll. 2013. Workers with a chronic disease and work disability: problems and solutions. *Bundesgesundheitsblatt-Gesund* 56(3):406–414. DOI: 10.1007/s00103-012-162111.

12 Work Ability in Workers with Mental Diseases

Halina Sienkiewicz-Jarosz
Institute of Psychiatry and Neurology

CONTENTS

INTRODUCTION

Industrialized countries are confronted with significant demographic changes as a consequence of increased life expectancy. According to the latest Eurostat employment statistics, the European Union (EU) employment rate for people aged 20–64 years reached 73.1% in 2018. Projections for employment in the EU going forward are far from good, as an increase from 25.9% in 2010 to 50.2% in 2050 in the old-age dependency ratio (the ratio between the number of persons aged 65 years and above and the number of persons aged between 15 and 64 years, expressed per 100 persons of working age (15–64)), has been predicted. In Poland, that ratio has grown from 18.9 in 2008 to 25.3 in 2018 (EUROSTAT 2018; OECD/EU 2018). This ageing of the population generates challenges for the employers, health organization systems and insurance, and it provokes studies designed to assess the potential risk factors for premature work exit. National authorities are trying to promote work ability throughout the working life and the optimal use of human resources. In the context of work ability, the increasing prevalence of mental diseases seems to be very important. According to Global Burden of Disease (GBD) 2015 Disease and Injury Incidence and Prevalence study (Vos et al. 2016), in middle-aged adults, mental health disorders, especially depression, are the highest-ranking of causes of disability. The burden of mental and substance use disorders may be responsible for early retirement (van Rijn et al. 2014; Lahelma et al. 2015; Hiilamo et al. 2019). Mental and other disorders associated with increasing age might limit the capacities of the ageing workforce. On the other hand, adverse psychosocial working conditions contribute to poor mental health among workers and increase the risk of mental diseases (Laine et al. 2014). In Poland, about 1.5 million people need ambulatory psychiatric help, and zmiana na: almost 200,000 are treated in psychiatric wards every year (Cybula-Fujiwara et al. 2015). People with mental disorders are at higher risk of exclusion from the job market. This can be linked to employers' attitudes and their reluctance to engage mentally disabled people, the impact of mental disorders on employment through decreased productivity, increased risk of accidents, absenteeism and presenteeism. It should be noted, on the other hand, that professional activity might play a role in the stabilization of the mental state and disease recovery (Modini et al. 2016a, 2016b). This chapter presents current data regarding the impact of selected major mental illnesses on work ability and the results of a survey on the prevalence of depression in professionally active people in Poland.

THE GENERAL IMPACT AND PREVALENCE OF MENTAL HEALTH PROBLEMS IN THE ADULT POPULATION OF THE EU AND POLAND

There is a bidirectional relationship between mental health and employment. Workers with mental illnesses may increase costs through increased absence, sick pay, the need to hire temporary replacements, and health and safety issues (Hilton et al. 2009). On the other hand, due to mental illnesses, adult workers may leave the labour market and young people may never enter it. The Organization for Economic Co-operation and Development (OECD) reports that mental disorders are the main cause of disability benefit claims in Western countries (OECD/EC 2018), although the full costs of mental diseases for the individuals, employers and societies are very difficult to estimate.

The latest statistics of the Institute for Health Metrics and Evaluation (IHME), an independent global health research centre at the University of Washington, are alarming. More than one in six people across the EU countries are affected by mental health problems. In 2016, the mental disease prevalence rate in the EU was 17.3%, indicating that nearly 84 million people were affected. The most common mental problems include anxiety disorders affecting 25 million people (5.4% of the population), next depressive disorders, with an estimated prevalence of 21 million (or 4.5% of the population), and drug and alcohol use disorders affecting 11 million people usunąć (2.4%). Bipolar disorders and schizophrenic disorders affect another 5 million (1.0% of the population) and 1.5 million people (0.3%), respectively (OECD/EU 2018).

In Poland, a survey conducted among a representative sample of 10,081 people aged 18–64 usunąć showed the lifetime prevalence of common mental disorders diagnosed in accordance with the DSM-IV classification as alcohol abuse (10.9%) of study group, significantly more common in men (18.6%) than in women (3.3%), and alcohol dependence confirmed in 2.2% of respondents. Psychoactive substance abuse was confirmed in 1.3% (Kiejna et al. 2015). The second most common mental health problems were panic attacks with the prevalence of 6.5% and depression, with major depressive disorder in 3.0% and minor depressive disorder in 0.4% of the respondents. Generalized anxiety disorder (GAD) was observed in 1.1% and lifetime dysthymia in 0.6% of the respondents. The prevalence of common mental disorders observed in Poland seems to be lower than in other EU countries (OECD/EU 2018).

COMMON MENTAL DISORDERS – EPIDEMIOLOGY AND CHARACTERISTICS IN THE CONTEXT OF WORK ABILITY

SCHIZOPHRENIA

Eurostat data show that the prevalence of schizophrenia in 2016 was 132.3 per 100,000 inhabitants of Poland. The risk of developing the disease is the same for both sexes, but it usually begins earlier in men: the onset of symptoms is observed between 15 and 24 years of age in men and between 25 and 34 years of age in women. It is thought that every hundredth person in the population will develop schizophrenia during their lifetime (Saha et al. 2005).

The causes of schizophrenia are complex and not fully understood. Currently, biological, genetic, psychological, social and environmental factors are taken into account. The risk of developing the disease is higher in families in which first-degree relatives (e.g. parents, siblings) have been diagnosed with schizophrenia. The risk varies depending on which member of immediate family is sick. For example, if both biological parents have schizophrenia, the likelihood that their child will get it reaches as much as 46%. Diagnosis is based on the patient's history, symptoms and behaviour. There is no specific biochemical test for diagnosing schizophrenia (Möller 2018). The course of schizophrenic disorders is usually varied: about 45% of patients recover after one or more episodes, and full or partial remission may be observed, but a progression of the disease is rather usual. A diagnosis of schizophrenia should not be made in patients with severe depressive or manic symptoms. Moreover, schizophrenia should not be diagnosed in the presence of a severe brain disease, or in a situation of substance intoxication or withdrawal. Treatment includes non-pharmacological methods, e.g. cognitive-behavioural therapy and, above all, pharmacological therapy with first- and second-generation antipsychotic drugs, and in individual cases with depot antipsychotics (Khan et al. 2016; Sieradzki 2019; Tracz-Dal 2019).

Schizophrenia is not a very common disorder (lifetime prevalence of 0.3%–1.0%), but has a very strong impact on work abilities. Importantly, patients with schizophrenia, independently of country of residence, spend very long periods in hospital (Charlson et al. 2018; OECD/EU 2018). For example, according to one study, the mean lenght of stay in a hospital due to schizophrenia in Great Britain was over 100 days, while in Czech Republic it was over 70 days. Polish patients spend on average over 60 days in the hospital every year. People with schizophrenia are also at the highest risk of unemployment. Polish data show that after about 10 years post-diagnosis, only 19% of patients work (Kiejna et al. 2013). Age at onset affects the outcomes, especially the number of hospitalizations, prevalence of negative symptoms, number of relapses, and social and occupational functioning and global outcome (Immonen et al. 2017). A metaanalysis of randomized trials showed that supported employment approach, combined with interventions targeting the significant deficits associated with schizophrenia, may help to improve employment outcomes (Carmona et al. 2017; Charlson et al. 2019).

Anxiety Disorders

Anxiety is a psychophysiological condition characterized by fear if imminent danger and helplessness, accompanied by specific somatic symptoms resulting from the stimulation of the vegetative system. Biological and genetic factors play a role in the development of anxiety disorders (obsessive-compulsive disorder is observed in one-third of the relatives of suffers), but environmental (psychogenic and exogenous), social or cultural factors are also meaningful. A higher prevalence of anxiety disorders is observed in women and people living in big cities, and slightly higher rates occur in populations with a lower socio-economic status and in those who are divorced and widowed. The probability rate of developing an anxiety disorder over the course of a lifetime is about 15%, and the peak incidence of this disorder is noted between 24 and 44 years of age (Kiejna et al. 2015). According to the World Health Organization (WHO), in 2017 5.1% of the Polish population experienced anxiety disorders. Anxiety is one of the most common symptoms associated with depression, and usually indicates a more serious clinical course and a greater risk of treatment discontinuation. The therapy for anxiety disorders includes psychological and pharmacological interventions (usually short-term benzodiazepines or antidepressants, especially serotonin reuptake inhibitors), optimally and individually selected for the patient (Bandelow et al. 2017).

In people with anxiety, increased absenteeism (days out of work) and presenteeism (days with lost productivity) rates have been observed. In the workplace, anxiety disorders may manifest as fatigue, difficulty in concentrating, worrying, or lack of self-confidence (Hilton et al. 2009). Workplace-related anxiety, including phobia, particularly when a patient has nonspecific somatic complaints should be also taken into account. A lower productivity level is a possible consequence, especially in GAD (Ivandic et al. 2017).

Sleep Disorders

People with sleep disorders usually complain of unsatisfactory sleep in terms of its quality, length or appropriate sleeping time along with problems such as difficulties with falling or staying asleep. Periodic sleep problems are found in 40%–50% of the population, 1.5 times more often in women. Sleep disorders may be classified into (i) secondary disorders in which sleep problems, that usually represent symptoms of somatic or psychological diseases, e.g. excessive drowsiness caused by a medical condition, night leg cramps, snoring, sleep apnoea and (ii) idiopathic (primary) sleep disorders not related to those listed above, e.g. insomnia disorder, hypersomnolence disorder, narcolepsy, obstructive sleep apnoea hypopnea syndrome, central sleep apnoea syndrome and parasomnias. Circadian rhythm sleep disorders include many different problems with accelerated, irregular or delayed sleep phase syndromes. As a result, disorders of sleep are accompanied by a feeling of fatigue and worse functioning during the day. Persistent sleep disorders (both insomnia and excessive sleepiness) pose a significant risk of psychiatric disorders as well as abuse of and addiction to psychoactive substances. Insomnia is often a symptom of other disorders, i.e. mental disorders in 50%-60% of cases (mainly affective and anxiety disorders), addiction in 10%-20% of cases (mainly drugs and alcohol abuse) and somatic diseases in 10%–30% of patients. Taking all of the above into account, the differential diagnosis of sleep disorders requires a multidimensional approach. The therapy for insomnia, the most common sleep problem, includes cognitive-behavioural therapy and pharmacological treatment (hypnotics, Z-drugs, sedatives from the benzodiazepine group, antidepressants, antipsychotics, antihistamines and melatonin) (Riemann et al. 2017).

According to Lallukka and Kronholm (2016), work duties can influence the quality of sleep and can interfere with the patient's personal chronotype. Sleep deprivation may be a consequence of high work demands and employer expectations disproportionate to a worker's capabilities. Moreover, stressful events in a workplace and interpersonal problems may be a cause of anxiety, which can additionally lead to a deterioration of the sleep quality over time. The consequences of worse sleep quality as regards work performance are indisputable. Whitney et al. (2015) have shown that sleep deprivation strongly impairs decision-making involving uncertainty and unexpected change. Sleep loss may induce cognitive impairment, which can even be dangerous during emergency response, disaster management, military operations and other dynamic real-world settings with uncertain outcomes and imperfect information. Long-term insomnia may cause adverse health outcomes; therefore, it should be analysed in a wider context as an important issue affecting public health and work ability (Lallukka and Kronholm 2016).

OCCUPATIONAL BURNOUT

Occupational burnout syndrome occurs when work ceases to be a source of worker's satisfaction and is accompanied by a lack of motivation. The worker feels overworked and dissatisfied with the activity he previously enjoyed. The causes of burnout syndrome can be divided into three groups: (1) individual factors (women, younger and better educated people are at risk), (2) interpersonal factors (worker-employer and worker-co-workers communication: unclear instructions, no dialogue, no impact) and (3) organizational factors (including work overload or unsatisfactory salary). According to Maslach et al. (1996), burnout syndrome can be characterized by emotional exhaustion, depersonalization (treating clients/students and colleagues in a cynical way) and reduced feelings of work-related personal accomplishment. Emotional exhaustion is a feeling of emptiness and an outflow of strength caused by excessive psychological and emotional requirements (or unrealistic requirements as against the worker's abilities). Symptoms of this condition include fatigue, frequent headaches, insomnia, sudden hot flushes, increased blood pressure, a sense of weakness, decreased body resistance, mood swings and even low-grade fever. Depersonalization means indifference to one's own needs and others' emotions, a sense of senselessness, cynicism towards others, reduction of sensitivity towards others and even voluntary exclusion from social life. The last symptom results from a reduction in the worker's assesment of the value of their own achievements and a sense of wasting time and effort at their workplace (Maslach and Jackson 1981; Bańkowska 2018). Negative attitudes outweigh the satisfaction with work, fear of the future begins and, as a consequence, a strong sense of being harm and guilt arise, generating further health problems. The WHO emphasizes that in order to be considered as burnout syndrome, these symptoms must occur in the context of work and not in other areas of life (WHO 2019). However, their consequences affect areas of family life and disrupt contacts with friends.

In general, the problem of burnout is still underestimated. According to most recent work by Lastovkova et al. (2018), the benefits through the social insurance to people with burnout syndrome in Europe are available in five countries only.

DEPRESSION

Depression is manifested mainly in depressed mood, decreased energy, loss of interest and pleasure, sleep and appetite changes, fatigue and concentration problems. According to the WHO, in Poland in 2017 5.1% of the population experienced depressive disorders. Depression at different degrees of severity affects over 350 million people worldwide, leading to disability (WHO 2017). It has been reported that depressive disorders occur twice as frequently among women than man, though this gender gap may be due to a higher tendency in women to report mental health problems. The incidence of chronic depression increases with age, and it is estimated that approximately 20% of people aged 65 years and above in Poland may suffer from depressive symptoms (Osińska 2017; OECD/EU 2018). The increase of the prevalence of depression in old age can be partly explained by poor physical health, worse financial status and lack of social support. The WHO predicts that by 2030, depression in terms of the total global burden of non-communicable diseases will overtake cardiovascular disease and become the most common disease. Suicide attempts are the most serious complication of depression.

The aetiology of depression is multifactorial, with the important roles played by biochemical (disruption of neurotransmission of serotonin, noradrenaline and dopamine in the limbic and reticular system), genetic (family history of affective disorders), social (significant deterioration of economic status, loss of employment, social position, lack of social support) or psychological factors (e.g. negative experiences from childhood, mobbing or harassment). In 85% of cases, depression is accompanied by other health problems, predominantly chronic diseases, such as rheumatoid arthritis or diabetes, and also by substance abuse, including alcohol and medicines.

In the workplace, depressive symptoms may manifest with restlessness, irritability, nervousness, fatigue, passivity and aimlessness. Some complaints may regard somatic health, like headaches or stomach pain. The clinical picture of the disease also includes decreased concentration and attention, low self-esteem, guilt and low self-value, pessimistic, black vision of the future, suicidal thoughts and acts, sleep disturbance and decreased appetite (Marcus et al. 2012; Osińska 2017). People with depression can lose about 27 days per year as a consequence of sick leave and loss of productivity. Depression is the third most important cause of disability in adults. Persons with mental disorders, including depression, experience lower employment rates by about 15%–30% as well as long-term unemployment (McDaid et al. 2015). In EU countries, only about half of the population aged 25–64 years with chronic depression were in employment, compared to more than 75% among those who do not report chronic depression (OECD/EU 2018). Work performance in people with depression is poor. It has been shown that workers with depression can achieve an acceptable level of work performance with some effort, and that their risk of worse work performance is two times higher than that of the general population (Ivandic et al. 2017). Many European countries are taking action to prevent mental illness, especially depression, and to promote mental well-being in all age groups.

PREVALENCE OF DEPRESSIVE SYMPTOMS IN PROFESSIONALLY ACTIVE POLISH PEOPLE – RESULTS OF A QUESTIONNAIRE STUDY PERFORMED IN 2018

According to the report of the Polish Ministry of Family, Labor and Social Policy of 2016, the proportion of people aged 50–64 years in the labour market was 58.4% compared to 69.0 on average in other EU countries. The rate of professionally active people aged 55–64 years was 49.9% and 59.6% in Poland and EU, respectively. The inducement for professional deactivation of people in this age group is the possibility of receiving a retirement pension and continuing working, and also social reasons, e.g. family responsibilities, and health problems. Mental disorders are among the main reasons for the increasing rate of people with disabilities or on sick leave, early retirement, work exit or job change, mainly in the 50+ population. According to Social Insurance Institution (Zakład Ubezpieczeń Społecznych, ZUS), in 2013, 93.9 thousand Poles benefited from sick leave due to depression (within a total of over 5.38 million sick days) (ZUS 2013–2014). The average duration of sick leave in depression was 57 days. In 2016, mental disorders accounted for 8% of all days of sick leave, and the first and subsequent depressive episodes were ranked as the 10th place and 24th most common causes of the longest sick leave (5.85 million days of work absence).

The main goal of the project was to determine the prevalence of depressive disorders in professionally active people, depending on the nature of their work, and to determine the social, clinical and economic risk factors for depressive disorders among professionally active people depending on age, work performed and economic conditions.

The survey was carried out on a randomly selected group of 1795 working people based on the Central Statistical Office (Główny Urząd Statystyczny, GUS) data on professional groups. The study was conducted using the "face-to-face" method by professional interviewers (Pen and Paper Interview, PAPI, BST Company). The sample was selected using the layer-random method. For this purpose, the population of the working people surveyed was first divided proportionally into four independent layers (by industry, gender, age and province). In addition, Polish Classification of Economic Activity (Polska Klasyfikacja Działalności, PKD) 2007 sections were adopted, and other categories were mapped directly from the Central Statistical Office data (GUS 2017). After preparing proportionate layers using the random method, the respondents were selected for the study from specific employment sectors until they reached at least the assumed number of interviews (Table 12.1). We used the previously described structured questionnaire (Konopko et al. 2018). The study was approved by the Ethics Committee of the Institute of Psychiatry and Neurology. All respondents gave their informed consent to participate in the study. The statistical analysis was performed using Statistica (Statsoft, version 13).

TABLE 12.1

The Number of Predicted and Completed Interviews According to Employment Sector

Trade	Numbers in Population	%	Predicted Number of Interviews	Number of Completed Interviews
Public administration	644,739	4.3	65	119
Construction	872,807	5.8	87	147
Education	1,152,094	7.7	116	119
Trade	2,299,084	15.4	230	239
Other	1,362,302	9.1	136	214
Healthcare	865,686	5.8	87	96
Manufacturing	2,671,042	17.8	268	271
Mining and quarrying	151,251	1.0	15	73
Agriculture and forestry	2,390,935	16.0	240	256
Service activities	2,554,471	17.1	256	261
Total	1,496,4411	100.0	1500	1795

The mean age of the participants was 35.3 ± 11.6 (23–65); 53.3% of the sample were men. People aged 30–39 years accounted for 38.4% of the respondents, those aged 40 to 49 for 30.2%, and those aged 50 to 59 for 27.1%. Twenty-five percent declared that they lived in villages, 26.9% in towns with more than 5000 inhabitants and 38.0% in cities with more than 200 000 inhabitants. Forty percent declared different levels of university education.

The prevalence of chronic diseases in the sample was as follows: 24.8% of respondents had been treated for hypertension, followed by orthopaedic diseases and diabetes. Other diseases, such as heart, kidney and endocrine disorders, were reported by about 13%–15% of respondents; 37.8% of study group confirmed that they were currently smoking, which is consistent with the results of a large survey on the prevalence of smoking in Poland (Kaleta et al. 2012). Every fourth respondent (25.2) reported that they drank alcohol more often than once a week. As many as 35.5% of the respondents reported drinking alcohol two to four times a month.

Fifteen per cent of respondents had been treated for depressive disorder in the past, and 4.1% confirmed current treatment with antidepressants. About twenty-six percent had a family history of depression.

When asked about the family situation, almost half of the study group (48.3%) answered that it was quite good, and that conflicts between the respondent and relatives were rare. Every third respondent (32.7%) assessed his family situation as very good. Fifteen per cent evaluated their family situation as difficult with frequent situations of conflict.

The economic situation was self-evaluated as good or very good by almost 70% of respondents; and some financial problems were reported by more than 25%. Only 2.8% of respondents described their financial situation as bad; these respondents replied that they were struggling with great financial difficulties and that their financial resources were insufficient to cover their basic needs.

More than one quarter of the respondents had in the previous year been on sick leave for 10–20 days, whereas one-tenth of the sample (9.8%) declared longer sick leaves (20–60 days), and 1.9% of the respondents reported even longer spells (60–365 days).

The prevalence of depressive symptoms (Patient Health Questionnaire-9 or PHQ-9) was 32.1%, including 2.7% of respondents with severe and moderately severe depression, 7.5% of respondents with moderate symptoms of depression and 21.8% of respondents with mild depression. The severity of depressive symptoms was correlated with the duration of declared sick leave ($p < 0.0001$).

The data presented above are, as indicated, part of results of the study conducted in 2018 on professionally active people in Poland. In the study group, depressive symptoms were more prevalent compared to statistics of Health at a Glance: Europe 2018 for Poland (OECD/EC 2018), where our country was on the list of countries with the lowest prevalence of mental disorders (rates below 15%). It has been suggested that some of the cross-country differences may be due to different levels of awareness and stigmatization associated with mental illness. Other important factors include availability of mental health services and a widespread belief that it is better to simply avoid talking about mental illness (Munizza et al. 2013).

More severe depressive symptoms correlated with longer duration of the declared sick leaves taken during the previous year. This observation is congruent with the previous data (Cybula-Fujiwara et al. 2015). Moreover, depression is a strong risk factor for disability pension (Amiri and Behnezhad 2019). People with depression more frequently change their place of work and are at risk of losing their job, so the prevention and treatment strategies are arousing more and more interests, as well as methods dedicated to helping people who plan to return to work following depression (Cybula-Fujiwara et al. 2015; Nazarov et al. 2019). In Poland, it seems very important to increase the awareness of the role of mental health among health professionals. Physicians need to inquire about the impact of depression on work and monitor the impact of symptom reduction on the recovery of work ability. Moreover, much effort is needed to develop individually addressed work-focused interventions (Nazarov et al. 2019).

CONCLUSIONS

Mental disorders affect a very high proportion of global population and generate a substantial burden, which has increased during recent decades. Further research is needed to better understand their causes and dynamics, and to prepare effective preventive strategies and public health responses.

REFERENCES

Amiri, S., and S. Behnezhad. 2019. Depression and risk of disability pension: A systematic review and meta-analysis. *Int J Psychiatry Med* 2019:91217419837412. DOI: 10.1177/0091217419837412.

Bandelow, B., S. Michaelis, and D. Wedekind. 2017. Treatment of anxiety disorders. *Dialogues Clin Neurosci* 19(2):93–107.

Bańkowska, B. 2018. Wypalenie zawodowe. Dylematy wokół istoty zjawiska oraz jego pomiaru. *Polskie Forum Psychologiczne* 23(2):430–445.

Carmona, V. R., J. Gómez-Benito, T. B. Huedo-Medina, and J. E. Rojo. 2017. Employment outcomes for people with schizophrenia spectrum disorder: A meta-analysis of randomized controlled trials. *Int J Occup Med Environ Health* 30(3):345–366. DOI: 10.13075/ijomeh.1896.01074.

Charlson, F. J., A. J. Ferrari, D. F. Santomauro et al. 2018. Global epidemiology and burden of schizophrenia: Findings from the global burden of disease study 2016. *Schizophr Bull* 44(6):1195–1203. DOI: 10.1093/schbul/sby058.

Charlson, F. J., M. van Ommeren, A. Flaxman, J. Cornett, H. Whiteford, and S. Saxena. 2019. New WHO prevalence estimates of mental disorders in conflict settings: A systematic review and meta-analysis. *Lancet* 394(10194):240–248. DOI: 10.1016/S0140-6736(19)30934-1.

Cybula-Fujiwara, A., D. Merecz-Kot, J. Walusiak-Skorupa, A. Marcinkiewicz, and M. Wiszniewska. 2015. Pracownik z chorobą psychiczną – możliwości i bariery w pracy zawodowej. *Med Pr* 66(1):57–69. DOI: 10.13075/mp.5893.00173.

EUROSTAT. 2018. https://ec.europa.eu/eurostat/tgm/table.do?tab=table&init=1&language=en&pcode=tps00198&plugin=1 (accessed September 18, 2019).

Hiilamo, A., R. Shiri, A. Kouvonen et al. 2019. Common mental disorders and trajectories of work disability among midlife public sector employees – A 10-year follow-up study. *J Affect Disord* 247:66–72. DOI: 10.1016/j.jad.2018.12.127.

Hilton, M. F., P. A. Scuffham, J. Sheridan, C. M. Cleary, N. Vecchio, and H. A. Whiteford. 2009. The association between mental disorders and productivity in treated and untreated employees. *J Occup Environ Med* 51(9):996–1003. DOI: 10.1097/JOM.0b013e3181b2ea30.

http://www.pzp.umed.wroc.pl/pl/article/2019/9/4/309 (accessed September 18, 2019).

https://www.senat.gov.pl/gfx/senat/pl/senatopracowania/175/plik/ot-674_zdrowie_psychiczne.pdf. (accessed September 18, 2019).

Immonen, J., E. Jääskeläinen, H. Korpela, and J. Miettunen. 2017. Age at onset and the outcomes of schizophrenia: A systematic review and meta-analysis. *Early Interv Psychiatry* 11(6):453–460. DOI: 10.1111/eip.12412.

Ivandic, I., K. Kamenov, D. Rojas, G. Cerón, D. Nowak, and C. Sabariego. 2017. Determinants of work performance in workers with depression and anxiety: A cross-sectional study. *Int J Environ Res Public Health* 14(5):E466. DOI: 10.3390/ijerph14050466.

Kaleta, D., T. Makowiec-Dąbrowska, E. Dziankowska-Zaborszczyk, and A. Fronczak. 2012. Prevalence and socio-demographic correlates of daily cigarette smoking in Poland: Results from the Global Adult Tobacco Survey (2009–2010). *Int J Occup Med Environ Health* 25(2):126–136. DOI: 10.2478/S13382-012-0016-8.

Khan, A.Y., S. Salaria, M. Ovais, and G. D. Ide. 2016. Depot antipsychotics: Where do we stand? *Ann Clin Psychiatry* 28(4):289–298.

Kiejna, A., P. Piotrowski, and T. Adamowski, ed. 2013. *Schizofrenia: perspektywa społeczna, sytuacja w Polsce.* Warszawa: Polskie Towarzystwo Psychiatryczne: Fundacja Ochrony Zdrowia Psychicznego.

Kiejna, A., P. Piotrowski, T. Adamowski et al. 2015. The prevalence of common mental disorders in the population of adult Poles by sex and age structure – An EZOP Poland study. *Psychiatr Pol* 49(1):15–27. DOI: 10.12740/PP/30811.

Konopko, M., W. Jarosz, P. Bienkowski, H. Sienkiewicz-Jarosz. 2018. Work-related factors and depressive symptoms in firefighters – Preliminary data. *MATEC Web Conf* 247. DOI: 10.1051/matecconf/201824700065.

Lahelma, E., O. Pietiläinen, O. Rahkonen, and T. Lallukka. 2015. Common mental disorders and cause-specific disability retirement. *Occup Environ Med* 72(3):181–187. DOI: 10.1136/oemed-2014-102432.

Laine, H., P. Saastamoinen, J. Lahti, O. Rahkonen, and E. Lahelma. 2014. The associations between psychosocial working conditions and changes in common mental disorders: A follow-up study. *BMC Public Health* 14:588. DOI: 10.1186/1471-2458-14-588.

Lallukka, T., and E. Kronholm. 2016. The contribution of sleep quality and quantity to public health and work ability. *Eur J Public Health* 26(4):532. DOI: 10.1093/eurpub/ckw049.

Lastovkova, A., M. Carder, H. M. Rasmussen et al. 2018. Burnout syndrome as an occupational disease in the European Union: An exploratory study. *Ind Health* 56(2):160–165. DOI: 10.2486/indhealth.2017-0132.

Marcus, M., T. Yasamy, M. van Ommeren et al. 2012. Depression: A global public health concern. World Health Organization, Department of Mental Health and Substance Abuse. https://www.who.int/mental_health/management/depression/who_paper_depression_wfmh_2012.pdf (accessed September 18, 2019).

Maslach, C., and S. E. Jackson. 1981. The measurement of experienced burnout. *J Occup Behav* 2(2):99–113.

Maslach, C., S. E. Jackson, and M. P. Leiter. 1996. *MBI: The Maslach Burnout Inventory: Manual.* Palo Alto: Consulting Psychologists Press.

McDaid, D., M. Knapp, H. Medeiros, and the MHEEN Group. 2008. Employment and mental health: Assessing the economic impact and the case for intervention. http://eprints.lse.ac.uk/4236/1/MHEEN_policy_briefs_5_Employment(LSERO).pdf (accessed September 18, 2019).

Modini, M., L. Tan, B. Brinchmann et al. 2016a. Supported employment for people with severe mental illness: Systematic review and meta-analysis of the international evidence. *Br J Psychiatry* 209(1):14–22. DOI: 10.1192/bjp.bp.115.165092.

Modini, M., S. Joyce, A. Mykletun et al. 2016b. The mental health benefits of employment: Results of a systematic meta-review. *Australas Psychiatry* 24(4):331–336. DOI: 10.1177/1039856215618523.

Möller, H. J. 2018. Possibilities and limitations of DSM-5 in improving the classification and diagnosis of mental disorders. *Psychiatr Pol* 52(4):611–628. DOI: 10.12740/PP/91040.

Munizza, C., P. Argentero, A. Coppo et al. 2013. Public beliefs and attitudes towards depression in Italy: A national survey. *PLoS One* 8(5):e63806. DOI: 10.1371/journal.pone.0063806.

Nazarov, S., U. Manuwald, M. Leonardi et al. 2019. Chronic diseases and employment: Which interventions support the maintenance of work and return to work among workers with chronic illnesses? A systematic review. *Int J Environ Res Public Health* 16(10):E1864. DOI: 10.3390/ijerph16101864.

OECD/European Union. 2018. Health at a glance: Europe 2018: State of health in the EU cycle, OECD Publishing, Paris/European Union, Brussels. https://doi.org/10.1787/health_glance_eur-2018-en (accessed September 18, 2019).

Osińska, M. 2017. Depression – Civilization disease of the 21st century. *Geriatria* 11:123–129.

Riemann, D., C. Baglioni, C. Bassetti et al. 2017. European guideline for the diagnosis and treatment of insomnia. *J Sleep Res* 26(6):675–700. DOI: 10.1111/jsr.12594.

Saha, S., D. Chant, J. Welham, and J. McGrath. 2005. A systematic review of the prevalence of schizophrenia. *PLoS Med* 2(5):e141.

Sieradzki, A. 2019. Schizophrenia and its health consequences as a public health problem. *Piel Zdr Publ* 9(4):309–313.

Tracz-Dal, J. 2019. *Zdrowie psychiczne w Unii Europejskiej. Opracowania tematyczne OT–674.* Warszawa: Centrum Informacyjne Senatu. https://www.senat.gov.pl/gfx/senat/pl/senatopracowania/175/plik/ot-674_zdrowie_psychiczne.pdf

van Rijn, R. M., S. J. W. Robroek, S. Brouwer, and A. Burdorf. 2014. Influence of poor heal h on exit from paid employment: A systematic review. *Occup Environ Med* 71:295–301.

Vos, T., C. Allen, M. Arora et al. 2016. Global, regional, and national incidence, prevalence, and years lived with disability for 310 diseases and injuries, 1990–2015: A systematic analysis for the Global Burden of Disease Study 2015. *Lancet* 388(10053):1545–1602. DOI: 10.1016/S0140-6736(16)31678-6.

Whitney, P., J. M. Hinson, M. L. Jackson, and H. P. Van Dongen. 2015. Feedback blunting: Total sleep deprivation impairs decision making that requires updating based on feedback. *Sleep* 38(5):745–754. DOI: 10.5665/sleep.4668.

WHO [World Health Organization]. 2017. *Depression and Other Common Mental Disorders: Global Health Estimates*. Geneva: World Health Organization. https://apps.who.int/iris/bitstream/handle/10665/254610/WHO-MSD-MER-2017.2-eng.pdf (accessed September 18, 2019).

WHO [World Health Organization]. 2019. Burn-out an "occupational phenomenon": International classification of diseases. https://www.who.int/mental_health/evidence/burn-out/en (accessed September 18, 2019).

ZUS [Zakład Ubezpieczeń Społecznych]. 2013–2014. https://www.zus.pl/baza-wiedzy/statystyka/rocznik-statystyczny-ubezpieczen-spolecznych (accessed September 18, 2019).

13 Activities Supporting Work Ability in Workers with Chronic Diseases

Joanna Bugajska
Central Institute for Labour Protection –
National Research Institute

CONTENTS

INTRODUCTION

Chronic diseases substantially contribute to limitations in work performance and thus work ability. The Eurostat data of 2017 reveal the prevalence of chronic diseases in about 30% of the European population. As compared to the data of 2010, the prevalence of chronic diseases increased by 6% (from 24% to 30%). An additional prevalence growth was noted during this period in occupationally active persons, from 19% to 28% (Llave et al. 2019). Chronic diseases progressively limit patients' everyday life functioning including work performance. The multifactor aetiology of many chronic diseases, including an important role of work-related factors and lifestyle in disease development, indicates that modification of these factors is one of the most important actions supporting employment in persons suffering from chronic diseases. This chapter presents an overview of the actions aimed at healthy lifestyle promotion and workplace adjustment considering the needs of workers with osteoarthritis, selected diseases of the cardiovascular system (coronary heart disease and arterial hypertension) and diabetes.

OSTEOARTHRITIS

As a rule, the first signs and symptoms of osteoarthritis occur in people between 40 and 60 years of age. They may affect one, several or (less often) multiple joints, and include joint tenderness, especially during movements and later also at rest and at night. The basic symptoms of osteoarthritis include pain and joint stiffness.

The later symptoms include bone deformities (bone spurs formed around the affected joint) and mobility limitation. The estimated prevalence of osteoarthritis is 9.6% and 18% in the male and female population over 60 years of age, respectively. In 80% of the population, it is manifested by a limited mobility and in 25% of the limitation of usual daily life activity performance is observed (WHO 2020).

The risk factors for osteoarthritis include (1) modifiable factors such as weakness of muscles surrounding the joint, joint structure deformities, either inborn or acquired (due to injuries), lack of physical activity, practicing sports involving exposure to excessive overload of joints and injuries, such as weightlifting, football or tennis; and (2) non-modifiable factors, which cannot be changed or are very difficult to modify them, such as age, sex (female) and genetic factors.

The risk factors for osteoarthritis also include exposure to multiple factors in the working environment, such as assuming an awkward body posture for a long period of time, hand-arm vibration and whole body vibration, lifting/moving heavy loads or repetitive movements.

LIFESTYLE RECOMMENDATIONS FOR WORKERS WITH OSTEOARTHRITIS

Actions aimed at lifestyle changes, including changes in dietary habits or participation in physical activities, contribute to pain reduction in patients with osteoarthritis, improving their social functioning and occupational activity performance. The aim of lifestyle modification in patients with osteoarthritis is, first of all, body mass normalization and minimization of other risk factors for joint degeneration (e.g. osteoporosis).

Diet

In patients with osteoarthritis, a proper dietary regimen should ensure maintaining a normal body mass. During diet planning, the recommendations related to comorbidities, especially diabetes and different kinds of lipid metabolism disorders, should be taken into consideration. The reduction of excessive body mass will be achieved by limiting calorie consumption and modifying diet composition. The goal of this chapter is to elucidate this issue.

REGULAR PHYSICAL ACTIVITY

Adequate physical activity is an important factor preventing osteoarthritis, helping people achieve and maintain a normal body mass and improve their general fitness and physical capacity levels. Besides, a regular involvement in physical activity, which is essential in case of chronic diseases, also improves the patient's mental comfort. The best effects can be obtained through participation in general rehabilitation gymnastics and non-weight-bearing exercises. Rational dosing of physical exercise engaging different parts of muscles, without undesired weight bearing, increases elasticity of periarticular structures and muscle strength, stabilizing this way certain joints.

Discussions focused on the advantages of physical activity are essential, especially for the Polish population. The results of the research conducted in over 200 occupationally active individuals with osteoarthritis indicate that only 50% of this population regularly, at least once a week, participate in a so-called active recovery workout including quick walks, nordic walking or biking, while as many as over 36% do it less often than once a month, very rarely or not at all (Bugajska and Księżopolska-Orłowska 2017).

WORKPLACE ACCOMMODATIONS FOR WORKERS WITH OSTEOARTHRITIS

Chances for getting a job or continuation of occupational activities in persons with osteoarthritis depend of the clinical course (periods of stability and aggravation) and location of the joints exposed to the degenerative process. During periods of stability when the symptoms are not so intense, there are no limitations in mental or physical work requiring little physical effort. During the periods of aggravation, when the increased sensitivity to pain and joint stiffness are often accompanied by other symptoms, e.g. accumulation of effusive fluid in the joints, it is recommended to limit work-related activities.

The activities that should be limited in persons with osteoarthritis include

- Hard and very hard physical work
- Working in exposure to substantial static load
- Activities that require assuming an awkward body posture
- Carrying heavy loads or people
- Working in exposure to cold and moist microclimate
- Working in exposure to hand-arm or whole body vibrations.

CORONARY HEART DISEASE AND HYPERTENSION

Cardiovascular diseases (CVDs) are the number 1 cause of death globally: more people die annually from CVDs than from any other cause. An estimated 17.9 million people died from CVDs in 2016, representing 31% of all global deaths. Of these deaths, 85% are due to heart attack and stroke. Most CVDs can be prevented by addressing behavioural risk factors such as tobacco use, unhealthy diet and obesity, physical inactivity and alcohol abuse using population-wide strategies (WHO 2017). CVDs manifest by numerous symptoms and complications (including stroke, myocardial infarction, or cardiac insufficiency) which lead to temporary or permanent inability to work. The most frequent morbid entities among the CVDs include coronary heart disease and arterial hypertension. Coronary heart disease usually affects economically active people, more often men aged 40–55 years, and among the older population (>55 years of age), the prevalence of this disease is similar in men and women.

Arterial hypertension is a very common disease, and potentially, it is most easily diagnosed. At the same time, it is the factor significantly increasing the risk of heart attack, cerebrovascular stroke, kidney failure and other diseases. In 2015, every fourth man and every fifth women presented with hypertension (WHO 2017). Both the exposure to adverse environmental factors and the tendency to prolong the time of occupational activity, as well as the continuous extension of life span in EU member states, lead to further growth in occupationally active population with circulatory disorders, especially coronary heart disease and arterial hypertension. The multiple actions undertaken to extend the length of employment in this population are obviously focused on the adjustment of job-related activities, but also on promotion of the so-called healthy lifestyle, as part of CVD prophylaxis.

LIFESTYLE RECOMMENDATIONS FOR WORKERS WITH CORONARY HEART DISEASE AND ARTERIAL HYPERTENSION

The actions focused on lifestyle changes including dietary habits, tobacco smoking, alcohol consumption or physical activity improve the efficacy of pharmacological treatment in patients with coronary heart disease and arterial hypertension and decrease the risk of cardiovascular complications in the course of the disease. The aim of actions focused on lifestyle changes in patients with CVDs is (1) to decrease blood cholesterol level, (2) to decrease blood triglyceride level, (3) to normalize the body mass, (4) to reduce the obesity including abdominal fat, (5) to provide pharmacological support and (6) to prevent other diseases, e.g. diabetes.

The main trends in lifestyle change presented next will be both the basis of CVD prevention and actions supporting pharmacotherapy in patients with these diseases.

Diet

Good nutrition prevents overweight and obesity and therefore excessive loading of the cardiovascular system and the development of other diseases (including osteoporosis, diabetes and some types of cancer). The relationship between nutrition and the risk of CVDs involves the effect of diet on such risk factors as cholesterol level, blood pressure and body weight. The dietary components which are particularly important from the point of view of cardiovascular risk include fatty acids, cholesterol and mineral components such as sodium and potassium.

Reduction of excessive body mass should be achieved by restriction of calorie content and diet composition, including limited salt and alcohol consumption, and tobacco smoking.

Physical Activity

Adequate physical activity is an important factor in prevention and non-pharmacological treatment of CVDs. Increased participation in physical activity also helps reduce overweight, improve overall physical fitness levels and decrease the risk of death due to CVDs. It should be recommended for patients with arterial hypertension to do at least 30–45-minute moderate aerobic exercise (running, quick walking, biking or swimming) within 5–7 days weekly. The type of physical activity should be adjusted to age, concomitant diseases and patients' preferences. Isometric exercises involving carrying heavy loads are not recommended. Patients with coronary heart disease should consult a physician prior to participation in any form of intense physical activity.

Discussing the advantages of physical activity is presently a very urgent problem. The results obtained in the population of over 300 occupationally active people with CVDs indicate that only about 55% of this population regularly, at least once a week, participate in physical activity (quick walking, *nordic walking*, biking) and as many as 30% participate in physical activity less often than once a month, which is very rarely, or never (Bugajska and Tyszkiewicz 2016).

WORKPLACE ACCOMMODATIONS FOR WORKERS WITH CORONARY HEART DISEASE AND HYPERTENSION

Multifactor aetiology is typical for CVDs. The common cardiovascular risk factors such as high cholesterol level, diabetes, inappropriate diet, genetic determinants, overweight/obesity, physical inactivity, sex (male) and a low socioeconomic status are the causes of about 50% prevalence of CVDs. Another 50% of cases are due to about 200 different factors including occupational factors (Pearson et al. 2003; Bortkiewicz 2011; Ezzati et al. 2002).

The major occupational factors responsible for CVDs include high physical (static and dynamic) load, toxic/chemical agents (e.g. carbon bisulphate, carbon oxide, nitroglycerin, lead), physical factors (noise, hot microclimate, cold microclimate, electromagnetic field), shift work, occupational burnout syndrome and work-related stress.

In people with CVDs, the decision to return to work requires taking into consideration disease progression, the degree of cardiovascular risk, the patient's well-being and the presence of the above-mentioned factors in the working environment.

Statistics show that a high percentage of Polish population after being diagnosed with CVD cease to return to work. In most of the cases, this fact is due to excessive caution and the lack of knowledge about work opportunities for people with CVDs (Kleniewska et al. 2012).

An enormous progress in diagnostics and treatment of CVDs has resulted in increased opportunities of return to work for people with such conditions, also those after sudden incidents (e.g. myocardial infarction).

Coronary Heart Disease

Patients with coronary heart disease, during the stable periods with no pain sensation while performing normal physical activities, are able to work, provided that their jobs do not require high-levels of physical effort or lifting/moving heavy loads. Caution should be exercised in case of exposure to physical workload or adverse microclimate. The return to work in people diagnosed with coronary heart disease, often first manifested by myocardial infarction, largely depends on the degree of disease progression, treatment approach and cardiovascular risk assessment.

Hypertension

Hypertensive patients with well-controlled blood pressure, and no diagnosed organ damage can continue their occupational activity. They can perform mental work, but also non-strenuous or medium hard work. Sometimes, however, it is necessary to change job, especially when it requires exposure to high mental load, stress and responsibility. Patients with arterial hypertension, working in exposure to environmental factors which are known to increase arterial blood pressure, such as carbon bisulphide or chrome and lead compounds, require special attention.

To sum up, the occupational activities which are not recommended for persons with CVDs include

- Hard and very hard physical work
- Exposure to high static load
- Work involving lifting and carrying heavy loads
- Working in exposure to physical factors including noise, especially impulse noise, hot microclimate, cold microclimate or electromagnetic fields
- Working in exposure to chemical agents such as organic solvents, carbon bisulphate, lead, carbon oxide and nitroglycerin
- Stress defined as high job requirements and low levels of control.

It is of note that CVDs as well as the medications used in treatment of these diseases may affect psychophysical fitness and contribute to movement coordination impairment or fainting. Therefore, under justifiable circumstances, it is advised to reflect on jobs requiring works at height, working with moving machinery or driving vehicles. It is also important to avoid over-normative working hours. Extra work at the same or another workplace increases exposure to physical load and stress.

Moreover, for people with a cardiac pacemaker, jobs involving exposure to electromagnetic field, such as electrical works, responsible for maintenance and repair of high voltage lines, as well as jobs involving operating medical equipment, e.g. MR scans and radio/television transmitters and high voltage lines, are contraindicated.

DIABETES

According to International Diabetes Federation data (IDF 2015), there were 59.8 million people with diabetes, accounting for 9.1% of the population aged 20–79 years in 2015. It is predicted that in 2040, the prevalence of diabetes will reach 71.1 million cases (10.7% of the population aged 20–79 years). These data indicate that diabetes is an enormously grave and continuously growing problem of contemporary communities since it affects a vast number of socially active people, including those continuing their education or being on the eve of getting a job suitable to their skills. Diabetes is a chronic metabolic disease resulting from carbohydrate metabolism disorders, often accompanied by lipid metabolism disturbances. There are several types of diabetes. Type 1 diabetes accounts for 10%–20% of all diagnosed cases of diabetes. It is due to the autoimmune process involving the occurrence of antibodies destroying beta-cells of the pancreas which produce insulin, a hormone responsible for glucose transport from blood to cells. In consequence, it causes growth of glucose blood level, being thus responsible for the onset of diabetes. Type 2 diabetes is a civilization-related disease, directly related to social factors. It accounts for the highest percentage of all types of diabetes, namely 80%. Its course is usually asymptomatic, and the disease is usually diagnosed at follow-up or is manifested by late complications such as vision impairment, myocardial infarction or cerebrovascular stroke. Other types of diabetes (e.g. steroid diabetes or gestational diabetes mellitus) account for about 3%–5% of all cases of diabetes.

Diabetes is a major cause of blindness, kidney failure, heart attacks, stroke and lower limb amputation. Healthy diet, regular physical activity, maintaining a normal body weight and avoiding tobacco use are ways to prevent or delay the onset of type 2 diabetes. The risk factors for diabetes are varied depending on the type. The risk factors for type 2 diabetes are best recognized. They include

- Age – the risk of disease development increases with age; type 2 diabetes usually affects the population after 45 years of age.
- Overweight or obesity – BMI value above 25, especially if accompanied by abdominal fat
- Improper diet
- Sedentary lifestyle
- Genetic factors
- Circulatory system diseases (heart diseases, elevated arterial blood pressure)
- Low levels of "good" cholesterol (HDL) or elevated triglyceride levels.

Based on the knowledge of the above-mentioned factors, we can conclude that healthy diet, regular physical activity, maintaining a normal body weight and avoiding tobacco use are ways to prevent or delay the onset of type 2 diabetes.

Complications in diabetes, especially damage to small and large blood vessels and the peripheral nerves, entail multiple health-related sequelae including vision impairment or blindness, kidney insufficiency, cerebrovascular stroke, myocardial infarction or limb amputation. Therefore, they lead to a substantial limitation in everyday functioning and thus limitation of occupational activities in patients with diabetes.

LIFESTYLE RECOMMENDATIONS FOR WORKERS WITH DIABETES

The main trends in lifestyle changes presented next are both the fundamentals of diabetes prevention and actions supporting pharmacological treatment in diabetics.

Diet in Diabetes

Patients with diabetes having a normal BMI values should maintain their body mass at this level while overweight and obese patients should reduce it. Planning the diet for diabetics, we should also consider the patient's individual preferences. It is necessary to choose products containing the desired nutrients, e.g. carbohydrate in the form of starch instead of monosaccharides and animal fat (butter) replacement with vegetable fats (olive oil). It is advised to divide the daily food ration into five to six meals, properly adjusted to the patient's lifestyle or job. It is worth remembering that good nutrition is critical in diabetes as it prevents the development of many other diseases including CVDs, osteoporosis or some types of cancer.

Systematic Physical Activity

Adequate physical activity is an important component in prevention and non-pharmacological treatment of diabetes. The goal of the actions aimed at increased participation in physical activity in patients with diabetes is to

- Improve metabolic control indices (24-hour glycaemic profile, lipid profile)
- Improve overall physical capacity
- Reach and maintain normal BMI
- Improve the patient's mental comfort.

At the time being, however, less than a half (40%) occupationally active people with diabetes regularly participate, at least once a week, in active recovery (e.g. quick walk, nordic walking, biking) and almost a half of this population rarely participates in such activities (Bernas et al. 2016).

Prior to and following exercise, it is advised to measure glycaemia levels, heart rate and arterial blood pressure. Sometimes, it is necessary to modify insulin doses and meal portions, depending on exercise intensity. The issues related to the form, duration and frequency of physical exercise performance should be consulted with a diabetologist.

Tobacco Smoking and Alcohol Consumption

In people with diabetes, tobacco smoking increases the risk of atherosclerosis even more than in healthy individuals. Besides, it increases the risk of myocardial infarction, cerebrovascular stroke or peripheral vascular diseases that often lead to limb amputation, vision loss and kidney insufficiency. Alcohol addiction in patients with diabetes, in turn, increases the risk of death due to acute hypoglycaemia and diabetic ketoacidosis/coma even by 50%.

The outcome of treatment in diabetic patients depends to a large extent on the patient's motivation and self-discipline. Education and motivation are the critical factors in successful treatment of patients with diabetes. Education should be focused both on the patient and his/her family, friends and coworkers.

WORKPLACE ACCOMMODATION FOR WORKERS WITH DIABETES

According to Polish Diabetic Society's recommendations, "The very fact of having diabetes cannot in itself constitute a reason for discrimination or unequal treatment. Occupational restrictions should be introduced after a careful analysis of individual circumstances and health status" (Marcinkiewicz and Walusiak-Skorupa 2018). Diabetes should to the lowest possible extent limit the patient's chances for achieving adequate education and employment according to the acquired skills.

However, research results show that such persons have problems with finding a satisfactory job. It turns out that 16% of patients with diabetes, being afraid of discrimination, hide their problem from the employer. The unemployment rate is twice as high in patients with diabetes than in the healthy population.

The patients with stabilized diabetes may have an unlimited access to education and suitable employment opportunities for a long time. There are multiple jobs, however, putting an excessive load on workers with diabetes.

There are two main reasons limiting employment opportunities in patients with diabetes. They include

- Hypoglycaemia (a very low level of blood sugar) and the associated consciousness impairment, entailing the risk of various signs and symptoms that may lead to various dangerous complications
- Late complications of diabetes, e.g. vision impairment or loss, adversely affecting work ability.

The risk of hypoglycaemia in workers with diabetes entails significant limitations and even contraindications for performing jobs connected with public safety, such as (Marcinkiewicz and Walusiak-Skorupa 2018):

- Professional driver (carrying passengers or goods, railroad and metro operators and taxi drivers)
- Uniformed and rescue services: armed forces (land troops, navy, aviation), police, fire brigade, shipping, prison service and licenced security guards
- Civil aviation, pilots and flight engineers, flight deck crew and air traffic controllers
- Especially dangerous jobs (work at heights, operating moving machinery, working in exposure to high temperatures, steel works, coal mines, and heavy traffic, and other jobs connected with high accident risk).

The complications occurring in the late stages of diabetes may also limit job opportunities since they impair the ability to perform specific work; such complications include vision impairment or loss, limiting the opportunities of getting multiple jobs requiring the utmost precision and vision clarity, such as watch-maker, jeweller and precision mechanic.

Additional attention and individual assessment of health-related occupational predispositions of persons with diabetes are recommended in cases of possible commencement or continuation of jobs requiring working in exposure to harmful and burdensome factors which may adversely affect the disease course. It particularly concerns jobs requiring working in exposure to high levels of physical effort or shift work, especially including night shifts. However, the list of jobs which are contraindicated for people with diabetes should not be extended only due to the lack of proper conditions and a friendly working environment for workers with diabetes, ensuring safe functioning of this working population.

TAKING MEDICATIONS AND SELF-CONTROL OF DIABETES AT A WORKPLACE

Taking medications, a properly balanced dietary regiment and self-control of diabetes at the workplace are important elements affecting work performance in workers with diabetes.

The therapy can be adjusted to worktime organization. This problem should be discussed with a physician. Each person suffering from diabetes should be allowed to control his/her blood sugar level. Each worker should expect an additional and unplanned physical effort or stress at work. In such cases, it is necessary to measure blood sugar level and adjust insulin doses and medication to this level. The diet also should be modified depending on physical effort intensity at work.

Worktime Organization

Regular working hours can facilitate self-control of glycaemia and the use of medications. Additional breaks should be considered for workers with diabetes so that they could take medications, have an extra meal or control glycaemia levels. Shift work or flexible working hours are not contraindicated, but they require better work organization both from the worker and the employer.

Workstation Adaptation

Each worker with diabetes should have enough private space for self-control of diabetes. It can be a cabinet protecting the equipment such as glucometer, an insulin pen, extra blades, cotton pads, strip tests and documentation in a form of self-control diary. Each worker should have enough room to determine blood glucose levels or insulin doses. As a rule, a well-educated patient knows how to prevent drops in blood sugar level. However, acute hypoglycaemia accompanied by the loss of consciousness may also happen. It is dangerous for the patient's health and even for his or her life. It is important to have a person among coworkers who is well trained in first aid and can properly react in cases of some undesirable events, such as hypo- and hyperglycaemia.

SUMMARY

Based on examples of actions undertaken to maintain the work ability of persons suffering from the four selected chronic diseases (osteoarthritis, coronary heart disease, arterial hypertension, diabetes) given in this chapter, it can be stated that these actions do not require significant financial resources. They are mainly focused on (1) adapting work organization and the psychosocial and physical work demands to the needs of the person with the disease, (2) building a good social climate to facilitate their functioning at work and (3) promoting a healthy lifestyle.

REFERENCES

Bernas, M. W., J. Bugajska, and E. Łastowiecka-Moras. 2016. *Warunki i organizacja pracy dla osób z cukrzycą. Poradnik.* Warszawa: CIOP-PIB.
Bortkiewicz, A., ed. 2011. *Choroby układu krążenia w aspekcie pracy zawodowej. Poradnik dla lekarzy.* Łódź: Oficyna Wydawnicza Instytutu Medycyny Pracy im. prof. J. Nofera.
Bugajska, J., and J. Tyszkiewicz. 2016. *Warunki i organizacja pracy dla osób z chorobami układu sercowo-naczyniowego. Poradnik.* Warszawa: CIOP-PIB.
Bugajska, J., and K. Księżopolska-Orłowska. 2017. *Warunki i organizacja pracy dla osób z chorobą zwyrodnieniową stawów. Poradnik.* Warszawa: CIOP-PIB.
Ezzati, M., A. D. Lopez, A. Rodgers, S. Vander Hoom, C. J. Murray, and Comparative Risk Assessment Collaborating Group. 2002. Selected major risk factors and global and regional burden of disease. *Lancet* 360(9343):1347–1360.

IDF [International Diabetis Federation]. 2015. *IDF diabetes atlas.* 9th Edition. http://www. diabetesatlas.org/resources/2015-atlas.html. (accessed January 28, 2020).

Kleniewska, A., M. Ojrzanowski, A. Lipińska-Ojrzanowska, M. Wiszniewska, and J. Walusiak-Skorupa. 2012. Bariery w aktywizacji zawodowej osób z chorobami układu krążenia. [Barriers to professional activity among people with cardiovascular diseases]. *Med Pr* 63(1):105–115.

Llave, O. V., J. Vanderleyden, and T. Weber. 2019. Working conditions. How to respond to chronic health problems in workplace? https://www.eurofound.europa.eu/publications/policy-brief/2019/how-to-respond-to-chronic-health-problems-in-the-workplace (accessed January 28, 2020).

Marcinkiewicz, A., and J. Walusiak-Skorupa. 2018. Zalecenia dotyczące aktywności zawodowej chorych na cukrzycę. *In Zalecenia kliniczne dotyczące postępowania u chorych na cukrzycę. Stanowisko Polskiego Towarzystwa Diabetologicznego. Diabetologia Kliniczna* 4(1):69–70.

Pearson, T. A., T. L. Bazzarre, S. R. Daniels et al. 2003. American Heart Association guide for improving cardiovascular health at the community level: a statement for public health practitioners, healthcare providers, and health policy make the American Heart Association Expert Panel on Population Prevention Science. *Circulation* 107(4):645–651. DOI: 10.1161/cir.0b013e31828f8a94.

WHO [World Health Organization]. 2017. *Cardiovascular diseases (CVDs).* https://www.who.int/news-room/fact-sheets/detail/cardiovascular-diseases-(cvds) (accessed January 28, 2020).

WHO [World Health Organization]. 2020. *Chronic rheumatic conditions.* https://www.who.int/chp/topics/rheumatic/en (accessed January 28, 2020).

Part IV

Work Ability and Disabilities

14 Model of Work Ability Assessment Using ICF

Joanna Bugajska and Andrzej Najmiec
Central Institute for Labour Protection –
National Research Institute

Karol Pawlak
International Classification of Functioning,
Disability and Health Council in Poland

CONTENTS

INTRODUCTION

According to World Health Organization (WHO) definition, disability is "a complex phenomenon, reflecting the interaction between features of a person's body and features of the society in which he or she lives" (WHO 2019). In the light of this definition, disability should not be assessed only from the medical point of view; the psychosocial aspects of functioning with a disability should also be considered.

In 2001, WHO accepted International Classification of Functioning, Disability and Health (ICF) (WHO 2001). This classification allows for a coherent approach towards evaluation of a human being, and its goal is to develop a standard and uniform language, enabling description of health status and the related conditions. ICF comprises all the aspects of human functioning as well as some elements of physical and mental well-being which are important for human health (Bickenbach et al. 1999). ICF is a universal model referred to all people regardless of their health concerns, age, sex and culture. It provides a sizable, comprehensive and standardized description of human functioning and the limitations in this functioning; it also serves as information management tool. ICF is composed of two parts, and each of them has two components. Part 1 of the classification comprises "Functioning and Disability", and its first component refers to "Body Functions and Structures". The latter is related to the human body and includes two classifications: the first classification is related to body system functions, and the second one concerns body structure. The second component is "Activity and Participation" and comprises the domains related to functioning from the perspective of a single individual and the whole society. Part 2 comprises "Context Factors". The first component of this part is related to "Environmental Factors" having a significant impact on each element of human functioning. The second component is related to Personal Factors which have not yet been classified. Qualifiers have been introduced to classification in order to describe impairment of function and structure of the body, limitations in activity and participation and the extent to which an environmental factor can make human functioning easier or more difficult.

According to ICF, disability is perceived as a dynamic interaction between a disease, injury or lesion and the context factors. Functioning and disability are not only the result of a morbid condition, but they depend on numerous factors including personal and environmental factors. The level of functioning and actual disability may differ in persons with the same morbid condition, but also the level of functioning and subjective perception of the quality of life may be similar in persons suffering from different conditions.

ICF does not classify people, but it describes the circumstances of human functioning in everyday life. The description is coherent with environmental and personal factors. Diagnosis based only on International Statistical Classification of Diseases and Related Health Problems – Revision 11 (ICD-11) does not provide a sufficient information on the real level of one's functioning and possible difficulties whose identification should be the basis for the development of the goals of rehabilitation programs (WHO 2018). Furthermore, ICD-11-based diagnose does not allow to determine the dynamics of problems encountered by a person with a disability whose level of functioning during a long process of treatment and rehabilitation may be subject to significant changes while ICD code remains unchanged. In this case, ICF is complementary to ICD-11, and, thanks to this fact, the health issues are referred to human functioning at a physical level (body functions and structures) as well as individual (activity) and social (participation) levels.

Due to its comprehensive nature, ICF is quite a complex tool to be used in everyday practice. Given this fact, WHO have developed a series of ICF-based instruments, such as ICF-checklist or WHODAS questionnaire.

In 2010, ICF was translated into Polish, and since then, attempts have been made to put it into practice.

USING ICF IN WORK ABILITY ASSESSMENT

According to WHO definition, work is an important aspect of human life and one of the most powerful social health-related factors (WHO 2012). It is also widely associated with being an integral part of a society and economic self-sufficiency. For many people, work is a source of self-confidence, self-esteem and social status. Job loss or no chance for return to work is frequently connected with feeling of despair and decreased self-confidence. Being unemployed is also associated with an increased risk of death (Roelfs et al. 2011). The growing number of disability claims and disability pensions contributes to increased financial burden to social security systems. It results in increased consciousness of the necessity to take all possible measures to prevent incapacity to work, including early implementation of occupational rehabilitation. The lack of motivation or opportunities for finding a job is not a sufficient reason for obtaining a medical certificate of incapacity for work (de Boer et al. 2008). Such a procedure requires complex medical assessment, focused on both body structure damage and the patient's functioning in their setting. Current evaluation procedures indicate a high potential of ICF in determining functional ability in people with disabilities and work ability assessment as well as social benefits and income support (Anner et al. 2012).

DESCRIPTION OF THE MODEL OF WORK ABILITY ASSESSMENT BASED ON ICF

The model of work ability for people with disabilities using ICF is composed of four stages of diagnostic procedures aimed at the development of a categorial profile in the following domains: body functions, activities and participation, and environmental factors. Such procedures are necessary for work ability assessment and the development of recommendations and proposed interventions in the context of return to work. The scheme of the model is presented in Figure 14.1, and the four stages of interventions are presented next.

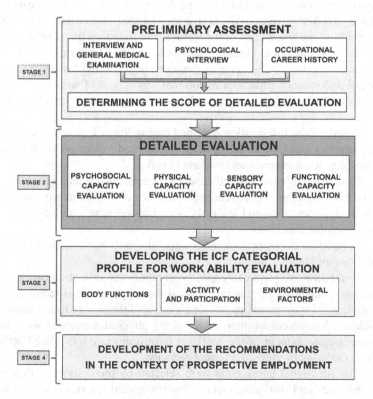

FIGURE 14.1 The scheme of work ability assessment model for people with disabilities by Bugajska, Najmiec, Pawlak 2018

Work ability assessment model for people with disabilities includes four stages of diagnostic procedures based on ICF, which are given as follows:

Stage 1

The first stage of work ability assessment in people with disabilities comprises general medical and psychological interviews, occupational history and general medical examination for making diagnosis according to ICD and according to the category of ICF structure. ICF structure categories are not added to the categorial profile of work ability assessment according to ICF since it would comprise all the above-mentioned categories which do not determine a general specifics of work ability, but only individual health status of a person with disability. It is necessary for work ability assessment to combine ICF with ICD. In the context of various diseases and injuries, a sole medical diagnosis may not ensure full conceptualization of health status and may not fully predict the patients' needs in terms of services or at the levels of individual treatment planning or healthy population policy. If we use only ICD, we may not acquire useful information necessary for planning and health management. Therefore, using ICF combined with ICD enables collection of data, providing a full picture of health and functioning in a coherent and comparable way. Based on medical diagnosis and interviews carried out during the first stage of evaluation, a team of specialists selects the tools and methods of the assessment of the features and functions that should be evaluated due to individual characteristics of a person with disability and his or her ability to work. The range of evaluated physical and functional, psychosocial and sensory capacities in complex assessment of work ability should be individually determined, depending on the type and degree of disability.

Stage 2

Based on the information obtained during Stage 1, psychosocial, physical, sensory and functional capacity is evaluated (using job-related simulations, e.g. work samples and baskets of task).

Evaluation of psychosocial capacity components reflecting work ability comprises evaluation of psychophysical well-being, personality traits, occupational preferences and interests, social competence, and cognitive and psychomotor capacities.

In justified cases, certain inborn qualities will be assessed (intellectual abilities and temperament).

Evaluation of physical capacity components reflecting work ability assumes diagnosis of general physical capacity, maximal strength capacity, range of movements in selected joints and the ability to maintain body balance. Measurement methods assessing physical aspects of work ability comprise assessment of the basic yet most important job-related activities from the point of view of physical capacity.

The functional parameters were selected by the authors of this model using the tool for diagnosing psychophysical capacity according to Functional Capacity Evaluation (FCE) (Soer et al. 2008). FCE is performed during job-related simulations and complemented by self-assessment of physical skills, and the experienced pain. A tolerance for highly repeatable activities, the imposed tempo and prolonged standing or positions with arms raised above shoulders, is also considered.

Determining the patient's sensory capacity, especially the capacity of the visual and auditory organs, in complex assessment of work ability in cases when medical assessment and interviews performed in the first stage of evaluation, indicates that limitations in sensory capacity might affect work ability.

Apart from quantitative assessment, the assessing specialists collect information based on observation of the behaviour of people with disabilities and case history based on the collected data, they determine the personality traits according to categorial profile of work ability assessment according to ICF.

Stage 3

During the third stage of work ability assessment, a team of specialists recodes the results, obtained from the scales of each evaluation method and tool into the categories of assessment in ICF qualifiers. Based on the recoded results and the conclusions from subject observation, a team of specialists develops an individual categorial profile including three categories, namely, body function, activity and participation, and environmental factors. The evaluation is based on cumulative information obtained during the first and the second stages of the following procedures: case history, patient's questionnaire, clinical and specialist evaluation and observation during the examinations and surveys.

Stage 4

The last stage of the model of work ability evaluation in persons with disabilities, based on ICF, involves development of recommendations for approaches to training, rehabilitation and occupational reorientation resulting directly from the patient's individual categorial profile reflecting work ability in persons with disabilities according to ICF. At the end of this stage, there is a meeting with the assessed participant including presentation of assessment results.

DIAGNOSTIC TOOLS AND METHODS ASSIGNED TO EACH ICF CODE (VERIFIED AFTER THE PILOT STUDY)

The methods and tools assigned to selected ICF codes, which enable information collection and/or objective assessment of the evaluated body functions, activity and participation as well as environmental factors, are presented in the Table 14.1.

RESULTS OF THE PILOT STUDY INVOLVING VERIFICATION OF THE MODEL

As part of a research project funded by the State Fund for the Rehabilitation of Persons with Disabilities and implemented in two centres specializing in comprehensive assessment of the competences and working capacity of persons with disabilities – Central Institute for Labour Protection – National Research Institute and Foundation for Human Resource Development, a research was conducted to

TABLE 14.1
Exemplary Diagnostic Tools and Methods Assigned to Selected ICF Codes

Code	Category	Diagnostic Tools and Methods
B114	Orientation functions	Observation Work samples, e.g. *sorting objects with hands,* *assembly simulation.*
B 117	Intellectual functions	Psychological tests: Wechsler, CFT 20-R Cattella (CFT 20-R – Cattell's Fluid Intelligence Test) Case history Observation
B 126	Temperament and personality functions	Psychological tests: (Neuroticism Extraversion Openness Five Factor Inventory NEO – FFI), Observation
B130	Vital energy and vitality functions	Questionnaire (Formal Characteristics of Behaviour – Temperament Questionnaire), Case history Observation
B1301	Motivation	Psychological tests: LMI – Achievement Motivation Inventory Observation Work samples: *sorting objects with hands,* *Assembly simulation.*
B140	Attention functions	Psychological tests: Attention and Perceptivity Test Work trials: *sorting objects with hands,* *assembly simulation* Baskets of task Case history Observation
B144	Memory functions	Psychological tests: Benton test, Rey–Osterrieth complex figure test, Wechsler Adult Intelligence Scale (WAIS) test Case history Baskets of task Observation
B152	Emotional functions	Interview, State-Trait Anxiety Inventory (STAI) Work samples: *assembly simulation,* *eye–hand–foot coordination* Baskets of task Observation

(Continued)

TABLE 14.1 *(Continued)*

Exemplary Diagnostic Tools and Methods Assigned to Selected ICF Codes

Code	Category	Diagnostic Tools and Methods
B280	Pain sensation	Medical examination
		Visual analog scale (VAS)
		Work samples: *whole body range of motion*
B455	Exercise tolerance functions	History (interview)
		6-minute test on a treadmill, Cooper test
		Work samples: *whole body range of motion*
		Observation
B710	Joint mobility functions	Medical examination
		Observation: step ladder, assessment of gait, Posture test
D175	Problem-solving	Baskets of task
		Case history
D240	Coping with stress and other mental burdens	Psychological tests: Composite Indicator of Systemic Stress (CISS)
		Case history (interview)
		Observation

verify the developed model. The research involved a pilot study using the model for verification of the developed diagnostic tools and methods. In each centre, 50 persons with disabilities were assessed. Additionally 20 people (from the total study group) were tested in both centres. The assessment was aimed at comparison the results; in total, the studied sample included 100 subjects with different types of disabilities (motor, hearing, vision or intellectual impairment) aged 18–60 years, with disability certificates. According to labour market status, the subjects were unemployed or job-seeking, or employed, but determined to change their jobs. The evaluation was carried out in three areas corresponding to three parts of ICF for Body Function category, Activity and Participation and Environmental Factors, essential from the point of view of work ability.

The assessment of Body Function was based on the presence of dysfunctions (according to ICF, called "impairments". The lowest rates of impairment were noted in the categories: hearing functions – 9%, thought functions – 10%, energy and drive functions – 13%, and memory functions – 15%.

The highest rates of impairment were noted in the categories: sensation of pain – 71%, vestibular functions – 43%, seeing functions – 47%, emotional functions – 38%, attention functions – 37% and exercise tolerance functions – 32%.

The assessment of activity function was based on problems with performance of a given activity. The lowest rating of problems was noted in the following categories:

- Formal relationships – 3%
- Basic interpersonal interactions – 5%
- Communicating with – receiving – nonverbal messages – 6%
- Communicating with – receiving – spoken messages – 6%

- Thinking – 6%
- Toileting – 6%
- Writing – 7%
- Using transportation – 7%
- Reading – 8%
- Conversation – 9%
- Dressing – 10%
- Complex interpersonal interactions – 12%
- Using communication devices and techniques – 13%.

The obtained results are very important in the context of undertaking occupational activities or occupational rehabilitation and vocational training. The above-mentioned activities are the basis for activating interventions; therefore, such rare problems with basic activity performance allow considering the outcome to be optimistic.

The highest number of problems was noted in the categories: vocational training – 69%, apprenticeship (work preparation) – 69%, driving – 57%, lifting and carrying objects – 48%, and maintaining a body position – 43%.

The results obtained from the research indicate the need of more intensive procedures aimed at increasing the participation in vocational training and courses among people with disabilities.

The values corresponding to physical and functional abilities may indicate some limitations in selected jobs and occupational activities; therefore, it is imperative to determine the occupational paths being in accordance with the individual's potential, based on work ability assessment.

The analysis of study results involved the comparison of the values describing two qualifiers for different categories in the area of Activity and Participation. The first one is *ability* qualifier describing the ability to undertake a given action by an individual, and the second one is *performance* qualifier describing the activities performed by an individual in his/her current setting.

The percentages obtained for part of the studied sample reporting no problems with specific task performance are presented in Table 14.2.

TABLE 14.2

Percentages Obtained from Part of the Studied Sample Reporting No Problems with Performance of Specific Tasks

	Type of Activity	Performance	Ability
1.	Carrying out daily routine	78	86
2.	Moving around using equipment	14	25
3.	Driving	38	54
4.	Looking after one's health	76	82
5.	Vocational training	24	93
6.	Higher education	27	75
7.	Apprenticeship (work preparation)	28	96
8.	Remunerative employment	61	94

These results indicate big differences between *abilities* and *performance*, particularly as regards the activities related to occupational activation of the subjects, including vocational training, higher education and vocational courses. This finding may indicate the presence of important personal of environmental factors, hindering performance in selected areas of functioning. As for Environmental Factors, the evaluation was focused on facilitators. The Environmental Factors with the highest rate of facilitators are presented in Table 14.3.

The above-mentioned areas with important facilitators can be divided into: social environment (family, friends, supporting professionals) and physical environment (products and technologies for communication, transport and facilitating everyday functioning). Social environment as a source of support is an important facilitator of tasks activating job search. These include support from families, friends as well as professionals cooperating with workers with disabilities. The high value corresponding to professionals cooperating with people with disabilities (for 71% of the sample, cooperation with these persons was a facilitator of different degree of intensity) indicates a high contribution of this occupational group to social and occupational activation of the study participants.

Access to products and technologies facilitating active functioning depends on the financial status and a given person' place of residence. People living in smaller towns and villages and those with a lower financial status have a limited access to such supporting factors. Economic self-sufficiency was noted in 70% of the sample, but the majority of participants lived in big cities where the opportunities for undertaking various forms of activity are higher. The verification procedures carried out in two research centres have shown a high usefulness of ICF for work ability assessment in people with disabilities. The recommended categorial profile was composed of 89 codes: 21 in Body Function area, 44 in Activity and Participation and 24 in Environmental Factors, based on Vocational Rehabilitation Core Set, can be regarded as a universal tool to be used by many specialists engaged in the process of work ability assessment (ICF 2010). Selected codes consider all abilities,

TABLE 14.3

Percentages Corresponding to Part of the Studied Sample Experiencing Facilitators

	Environmental Factors	Percentage of Participants Experiencing Facilitators
1.	Health-related professionals	71
2.	Products and technology for communication	68
3.	Immediate family	67
4.	Products and technology for personal use in daily living	66
5.	Transportation services, systems and policies	65
6.	Drugs	65
7.	Design, construction and building products and technology of buildings for public use	63
8.	Friends	61

possibilities and environmental determinants which are important from the point of view of work ability in people with different kinds of disabilities.

Complex assessment of work ability should contribute to increased activation of people with disabilities, focused on employment. Taking into account the information collected and evaluated by the team of specialists should provide us with a complete and objective picture of the individual's potential, encountered difficulties and different types of support.

In order to reach the desired goal, advantage should be taken of informative and emotional elements of support.

Work ability assessment is made by the interdisciplinary team of specialists having experience in: medicine, health psychology and occupational psychology, physiotherapy and biomechanics, occupational counselling and occupational rehabilitation planning. Each member of the diagnostic team is responsible for specific tasks so that a complex categorial profile of work ability would be developed. Following the evaluation, the whole team develops an agreed and final version of work ability categorial profile. The developed profile is the basis for recommendations suggesting specific actions necessary for a person with a disability to find a job. The choice of training directions, further rehabilitation and occupational reorientation, resulting from work ability assessment, is made with active participation of a person concerned, consulted and selected by this person before making a final decision.

CONCLUSIONS

Work ability assessment using ICF may be the first step to increase the activity oriented at job taking. Considering the wide spectrum of information obtained and assessed by a multidisciplinary team of specialists may reveal a full picture of both the potential of the study subjects and the encountered barriers. The assessments of functions, activity and participation, and environmental factors are interdependent. A complex evaluation of work ability enables them to detect these interdependencies. In our opinion, evaluation of work ability will allow people with disabilities to get to know their potential, increase their motivation for taking a job and change their attitude towards a disability.

REFERENCES

Anner, J., U. Schwegler, R. Kunz, B. Trezzini, and W. de Boer. 2012. Evaluation of work. Disability and the international classification of functioning, disability and health: What to expect and what not. *BMC Public Health* 12:470. DOI: 10.1186/1471-2458-12-470.
Bickenbach, J. E., S. Chatterji, E. M. Badley, and T. B. Üstün. 1999. Models of disablement, universalism and the international classification of impairments, disabilities and handicaps. *Soc Sci Med* 48(9):1173–1187. DOI: 10.1016/s0277-9536(98)00441-9.
Bugajska, J., Najmiec, A., Pawlak K. 2018. Applying of the International Classification of Functioning, Disabilities and Heath (ICF) for the Assessment of Work Ability. *Disabilities-issues, problems solutions* 29: 67-80.
de Boer, W. E. L., P. Donceel, S. Brage, M. Rus, and J. Willems. 2008. Medico-legal reasoning in disability assessment: A focus group and validation study. *BMC Public Health* 8:1–9. DOI: 10.1186/1471-2458-8-335.

ICF [International Classification of Functioning, Disability and Health]. 2010. ICF Core Sets for Vocational Rehabilitation. https://www.icf-research-branch.org/icf-core-sets-projects2/diverse-situations/icf-core-sets-for-vocational-rehabilitation (accessed January 28, 2020).

Roelfs, D. J., E. Shor, K. W. Davidson, and J. E. Schwartz. 2011. Losing life and livelihood: A systematic review and meta-analysis of unemployment and all-cause mortality. *Soc Sci Med* 72(6):840–854. DOI: 10.1016/j.socscimed.2011.01.005.

Soer, R., C. P. van der Schans, J. W. Groothoff, J. H. Geertzen, and M. F. Reneman. 2008. Towards consensus in operational definitions in functional capacity evaluation: A Delphi survey. *J Occup Rehabil* 18:389–400. DOI: 10.1007/s10926-008-9155-y.

WHO [World Health Organization]. 2001. International Classification of Functioning, Disability and Health (ICF). https://www.who.int/classifications/icf/en/ (accessed January 28, 2020).

WHO [World Health Organization]. 2012. Final Report of the World Conference on Social Determinants of Health: Rio de Janeiro, Brazil, 19–21 October 2011. https://www.who.int/sdhconference/en (accessed January 28, 2020).

WHO [World Health Organization]. 2018. International Statistical Classification of Diseases and Related Health Problems. Revision 11 (ICD-11). https://www.who.int/classifications/icd/en (accessed January 28, 2020).

WHO [World Health Organization]. 2019. Disabilities. https://www.who.int/topics/disabilities/en (accessed January 28, 2020).

15 Conditions Impacting on Work Ability in People with Motor Disabilities – Results of Research

Karolina Pawłowska-Cyprysiak
Central Institute for Labour Protection –
National Research Institute

CONTENTS

INTRODUCTION

Acquired disabilities require self-assessment of various spheres of life in persons with disabilities. It can be assumed that the type of disability will be a differentiating factor for this assessment, since persons with different types and degrees of disability differ in their ability to cope with problems and their potential capability to consciously assess their functional status.

Self-esteem is altered in people with acquired disabilities, and their relations and roles within the society and families are subject to significant changes. They suffer mentally, feel physical pain and have to cope with long-term treatment and gradually progressing limitations on their everyday functioning and activity performance (Bishop 2005). Body fitness impairment leads to work-related consequences, e.g. changes in occupational status. Persons with disabilities emphasize the adverse effect of their health status on their performance and continuation of occupational activities, and perceive disability as the factor hindering their chance for promotions or limiting their opportunities for getting a satisfactory job (Lipińska-Lokś 2007). Research also shows that acquired disability negatively affects the level of satisfaction with income, social life and leisure time in this population, yet it positively correlates with satisfaction of having ample free time (Powdthavee 2009).

Coping with limitations resulting from physical disability is part of adaptation process. It involves a deliberate strategy which is expected to lead to stress coping ability. The strategy may be personality-specific (in this case, it is identified with a relatively constant personality trait) and is reflected by specific behaviour when facing a stressful situation. It can also be circumstance-specific (based on implementation of stress-coping strategy under stressful conditions, depending on the type of stressor), yet it does not have to be in accordance with the patient's personal predispositions (Byra 2014). The results of surveys conducted in patients with spinal cord injury suggest that social and educational support provided for both people with disabilities and their families can enhance this process (DeSanto-Madeya 2009). Independent functioning after spinal cord injuries, in turn, and pain perception levels are the predictors of the perceived life satisfaction (Van Leeuwen et al. 2011).

Social and environmental factors play an equally important role in disability acceptance as internal factors (Olney et al. 2004). Occupational activity belongs to such factors. In people with disabilities, it directly affects the quality of life and life satisfaction (Pawłowska-Cyprysiak 2011, Pawłowska-Cyprysiak et al. 2013a, 2013b). Occupationally active individuals with disabilities give definitely higher scores to their family relations and report having the same status as other family members (occupationally passive individuals have reported being often treated on special terms), their self-rating health status is higher and they perceive themselves as fit for numerous activities. Occupationally active individuals spend more time learning and have a high self-esteem, whereas occupationally passive persons spend more time performing activities that require less effort, and their self-esteem is lower. These are usually people who have accepted their disabilities and earn their households. The occupationally passive persons with disabilities, in turn, perceive themselves and their health status as worse compared with others. Those who fail to accept their disability are prevailing in this group. Most of them cover their living expenses from disability benefits (Brzezińska et al. 2008).

For people with disabilities, work is a source of income and allows social integration, feeling normal and not different from other people. Such attitudes are important from the point of view of the motivation for return to work and of the health promotion (Saunders and Nedelec 2014).

RETURN TO WORK AFTER ACQUIRING A DISABILITY

There are multiple factors determining whether the person with an acquired motor disability will return to work or not. These include some barriers in the labour market (e.g. employers stereotypes about disability) (Vornholt et al. 2018) and individual factors, such as a decreased work ability, age when the disability was acquired (the persons who acquired a disability when they were younger and are characterized by a greater independence sooner return to work) (Lidal et al. 2007), assessment of health status (Lidal et al. 2007; Brzezińska et al. 2010) or the level of intrinsic motivation (Wolski 2013). Higher education levels and willingness for continuing education are the factors promoting the return to work in persons with acquired disabilities (Ottomanelli and Lind 2009). On the other hand, a fear of losing benefits is one of the reasons why people with acquired disabilities are reluctant to return to the labour market. The benefits assure regular cash flow and meet basic needs of people with disabilities and their families (Wolski 2010).

Adaptation of the worksite to disability is of particular importance. It enables people with disabilities to correct health-related limitations and promotes a more efficient and better functioning at work. Reasonable adjustments allow workers to take advantage of their skills and personal resources and guarantee safe functioning in the working environment (Zawieska 2014). Employers also suggest that worksite adaptation may encourage people with disabilities to return to work (Wainwright et al. 2013), highlighting the importance of rising awareness about disability at worksite and emphasizing the role of social support for people with disabilities (from co-workers and superiors) (Pawłowska-Cyprysiak 2015).

However, we shouldn't forget the importance of complex work ability assessment in determining individual competences of candidates and support in career choice to increase their chance of remaining in the labour market (Bugajska 2018).

WORK ABILITY

Self-rating of work ability is a result of the correlations between physical and mental work demands and the worker's potential, skills and health status (Tuomi et al. 1994). The study results presented by Miranda et al. (2010) indicate that multisite pain is a factor contributing to a decrease in work ability. We should remember, however, that workers' low self-rating of their health status is not always in accordance with their low self-rating of work ability.

The results of Health Survey 2000 (Koskinen et al. 2008) indicate that one-third of the respondents reporting poor health also report no limitations of their work ability. However, all chronic conditions contribute to reduction of work ability (the finding also confirmed by the results reported by El Fassi et al. (2013) where the result of logistic regression analysis indicates that the reported diseases reduced work ability). Mental disorders and coronary artery disease are the conditions having the strongest negative impact on work ability. Workers report depression and conditions affecting the neck and back as the main reasons for work ability impairment (Koskinen et al. 2008). The above finding is confirmed by the results presented by Ge et al. (2018), indicating a significantly decreased work ability in persons complaining of back, neck and shoulder pain. Tuomi et al. (1991) has also found that musculoskeletal disorders contribute to a decreased work ability. Such a decrease and the resulting premature exclusion from work-related activity and social lives in persons with musculoskeletal disorders is the consequence of overload syndromes (Bugajska et al. 2011).

In people with disabilities, work ability depends not only on body injuries, but also on numerous factors including intelligence level and participation in rehabilitation procedures that help regain self-sufficiency and improve functioning, the level and type of job-related skills, personal characteristics (self-evaluation, self-esteem, career aspirations, motivation to work, determination and persistence in goal achievement), being well prepared for the job on the open labour market and environmental factors (place of living, availability of transport, development and infrastructure of a given area, family relations). The cumulative effect of the aforementioned factors is defined as employability of a person with a disability (Majewski 2011). Given the fact that modification of the factors affecting work ability assessment is possible, it is justifiable to establish work ability determinants in occupationally active and inactive persons with acquired motor disabilities, in order to use suitable strategies to activate this group of people.

The goal of the research carried out between the years 2014 and 2016 was to indicate factors determining work ability in people with acquired motor disabilities.

METHODS

The survey was conducted using Paper and Pencil Interview (PAPI) technique, involving a direct interview using a paper questionnaire, which was filled by an experienced surveyor. The survey book included the following questionnaires.

PERSONAL QUESTIONNAIRE FOR WORKERS WITH ACQUIRED MOTOR DISABILITIES

This survey has been specifically developed for the research, based on the available literature pertaining to the studied issue and good practices in this field. It consists of 34 questions grouped into thematic blocks, including

- General information
- Information on the disability
- Information on the household
- Work history.

The main goal of the survey was to address the conditions of occupational reintegration in persons with acquired disabilities, to identify the motives for their return to the labour market as well as the participants' needs, in the process of occupational reintegration involving, in this case, flexible forms of employment.

PERSONAL QUESTIONNAIRE FOR THE UNEMPLOYED
WITH ACQUIRED MOTOR DISABILITIES

This is the questionnaire developed for the study, based on the analysis of the available literature and good practices in this field. It contains 32 questions grouped in topical blocks:

- General information
- Information on the disability
- Information on the household
- The causes of unemployment
- The potential for occupational reintegration.

The main goal of the questionnaire was to address the reasons for not getting a job offer and the actions that should be undertaken to encourage non-working participants to return to the labour market.

The Acceptance of Illness Scale (AIS) Developed by B. J. Felton, T. A. Revenson i G. A. Hinrichsen Z., Polish Version by Juczyński

This scale assesses illness acceptance. Each statement contains description of one of the eight adverse health-related consequences including lack of self-sufficiency, feeling dependent on someone or a decreased self-esteem. Fewer negative consequences reflect a higher level of the respondents' acceptance of their health status. Estimated Cronbach's α value, based on the results obtained from 138 patients with chronic pain, was 0.85. The stability of the scale, in turn, estimated based on the twofold survey conducted at a 4-week interval was 0.64 (Juczyński 2012).

Life Orientation Test (LOT-R) Developed by M. F. Scheier, Ch. S. Carver, M. W. Bridges, Polish Version Adapted by R. Poprawa and Z. Juczyński

This scale evaluates dispositional optimism (including optimistic thinking, allowing effective coping with problems in everyday life, a tendency to experiencing positive or negative emotions and the related life satisfaction or dissatisfaction). The estimated Cronbach's alpha value, based on the results obtained from 174 persons, was 0.76. The final result of the twofold LOT-R test was 0.43 (Juczyński 2012).

WAI – Work Ability Index Developed by Tuomi, Ilmarinen, Jahkola, Katajarinne Tulkki, Polish Version by J. Pokorski

This index is an instrument measuring work ability. Its numerical value is calculated based on the score obtained for each question. WAI values range from 7 to 49 where

- 7–27 points indicate poor work ability.
- 28–36 points indicate moderate work ability.
- 37–43 points indicate good work ability.
- 44–49 points indicate excellent work ability.

Additionally, the non-working participants were asked to assess their current ability to do any work as compared with their lifetime best (top form) in 0–10-point scale (WAI, question 1).

General Self-Efficiency Scale Developed by R. Schwarzer, M. Jerusalem, Adapted by Z. Juczyński

This scale assesses general self-beliefs related to coping with difficulties and obstacles. Owing to this scale, it is possible to predict actions and intentions in different areas of activity. The respondents' sense of self-efficacy affects the assessment of their resources in stressful situations. The estimated value of Cronbach's alpha, based on the results obtained from 174 participants, was 0.85. The final result of the twofold test was 0.78 (Juczyński 2012).

STUDY GROUP

The sample included 500 participants with acquired motor disabilities. The participants were selected using quota and judgmental sampling where quotas were determined based on the occupational status and gender.

STATISTICAL ANALYSIS

Statistical analysis of the obtained parameters was conducted using SPSS program.
It included

- Descriptive statistics, providing sample characteristics
- The independent samples t-tests to determine significance level of the
  between-group differences
- Stepwise regression analysis to determine work ability predictors in the
  studied sample.

RESULTS

SAMPLE DESCRIPTION

The sample comprised 500 participants including 50% of women. The mean age of
the participants was 43.7 years (SD = 10.9). The mean time of living with a disability was 3 years (SD = 7.3). Thirty-seven per cent of the sample lived in towns with
50,000–100,000 inhabitants. Most of the respondents (61%) were in relationships and
had children (62%). As for the educational level, the biggest group included people
with basic vocational education (28%) and secondary vocational education (24%).
    Considering the disabilities, the sample included

- 49% of participants with mild disabilities, 40% with moderate disabilities
  and 11% with severe disabilities.
- 38% with lower limb impairment, 26% with upper limb impairment, 31%
  with "other" types of motor impairment and 5% did not answer the question.

The most frequent causes of disabilities included

- Road accidents (41%) and past diseases (32%).
- The participants assessed their health status as quite good or average (69%),
  and reported having ailments affecting their everyday functioning (56%).

WORKING PEOPLE

Among 250 working participants, 49.6% were women. The mean age of the studied sample was 42.8 years (SD = 10.6); 42.8% of the respondents lived in towns with
50,000–100,000 inhabitants. As regards disabilities, the sample comprised the highest

percentage of participants with mild disabilities (69.6%) including acquired lower limb deficiencies (38%). The mean time of living with a disability was 6.4 years (SD = 6.92). The most frequent causes of disabilities included road accidents (41%), past diseases (32%) and accidents at work (19%). Sixty-six per cent of the respondents rated their health status as quite good or satisfactory, whereas 58.4% and 42% reported that their ailments mainly affected their household duties and occupational activities, respectively.

As regards employment, 70% of the participants were private sector workers or worked in the open labour market (68%). Most of the respondents reported doing the same jobs as before (60%), whereas 53.2% remained at their workplace after acquiring the disability.

According to 54% of the respondents, the workplace was not adapted to suit their needs resulting from disability (they mentioned that the furniture was not adequately adjusted to their disabilities and it was impossible to leave the place during breaks). The reasons included lack of financial support, commitment and willingness from the management, or obstacles.

Conversely, the respondents confirming that the employer adjusted the workplace to suit their disability-related needs mentioned such assistive solutions as special seating aids, grab bars in toilets, keeping physical activity to a minimum, providing adequate working instrumentation, widening doorways, moving the workplace to the ground floor, providing adequate lighting, ensuring transport to the workplace, ensuring rest in case of health-related problems, flexibility of working hours, and breaks at any time, when necessary.

The return to work after acquiring a motor disability was largely influenced by sense of security (job security) (40%), financial problems (35%), an opportunity to earn and to meet one's own and family's needs (35%), or feeling independent owing to employment (32%).

The two main factors helping the respondents to return to work included an opportunity for retraining and adjustment of the workplace for workers with disabilities (Figure 15.1).

Eighty-six per cent reported that they wouldn't like to change anything in their professional life. The remaining part of the respondents (68.6%) would like to change their workplace, whereas 48.6% would like to work in flexi-time system (usually to reduce their working week). Most of the respondents (80.8%) reported feeling accepted by their employers and co-workers.

According to 75.6% of the respondents, reported that the boss (superior) plays an important role in return to work. This role mainly involves communicating with the workers who acquired motor disabilities during their absence (83.6%), commitment from co-workers in creating a friendly working environment for their workmates with disabilities (83.2%), building proper relationships between the workers returning to work and their workmates (82%), showing concern for the health of workers with acquired motor disabilities (78.8%), meeting with the workers before their return to work (76.8%), analysis of disability-related needs to adjust the working environment to disability type (78.4%), and consultations with an occupational health physician (71.6%).

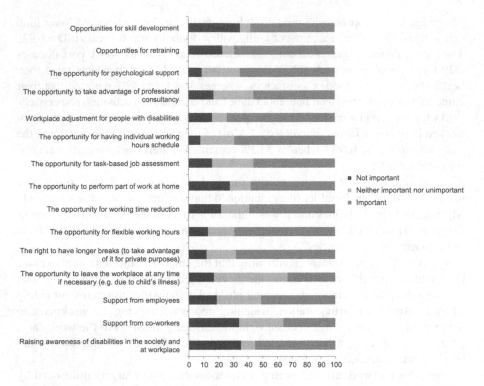

FIGURE 15.1 What do you think is most important for people with acquired disabilities when they plan to return to work?

Non-working Individuals

Fifty per cent of non-working group (N = 250) were women. The mean age in this group was 44.5 years (SD = 11.1); 30.8% of the participants lived in towns with 50,000–100,000 inhabitants, 62.4% reported being in a relationship and 29.6% had basic vocational education; 77.6% of the group lived with their families. Sixty-six per cent of the respondents reported having children. The mean time of living with a disability was 5.7 years (SD = 4.8). The most frequent reasons for the disability were road accidents (42%). The highest percentage of the studied sample reported having moderate degree disabilities (51%) including lower limb impairment (40%); 72.8% of the sample assessed their health status as relatively good or average, and 47% reported that their symptoms mainly affected their household duty performance.

Most of the unemployed participants (68%) assessed their financial health as relatively good or average. The main sources of income in this group were annuity (35%) and family members' earnings (32%).

Poor health was the main reason for not taking a job by the respondents (57%). Other important reasons for not taking a job included no job offers at or near the place of living (32%), incapacity for work (31%) and difficulties getting to work (26%).

The non-working respondents reported the factors that would help them return to work. These included adjustment of the workplace to workers with disabilities,

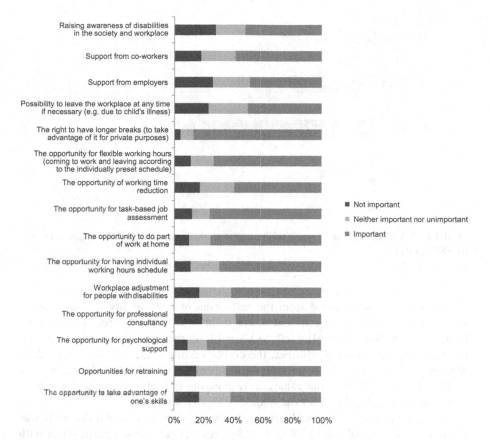

FIGURE 15.2 What do you think would help people with acquired disabilities to return to work?

support from the employers and opportunities for worktime reduction, doing at least part of work at home or working within flexi-time system (Figure 15.2).

An important role of the employer in return to work was indicated by 66.4% of non-working respondents. They believed that the main tasks of the employer included adjustment of the working environment to people with acquired disabilities (58%) and motivating the worker with acquired motor disability to work (56%).

WORK ABILITY OF WORKERS WITH ACQUIRED MOTOR DISABILITIES

The assessment of work ability in the studied sample of people with acquired disabilities was possible owing to the analysis of Work Ability Index (WAI) questionnaire, completed by the workers. For the non-working respondents, they were asked only one question related to their current work ability, compared with their lifetime best (WAI, question 1).

The self-rating of work ability revealed that 57% of the respondents rated it as moderate. Only 4% rated their work ability as excellent (Figure 15.3).

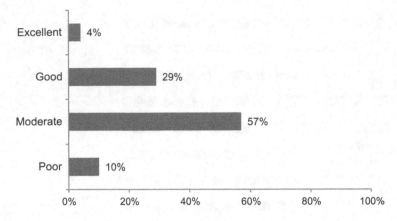

FIGURE 15.3 Assessment of work ability in the working participants with acquired motor disabilities (%).

The mean value obtained from the self-rating of work ability in the group of working participants with acquired motor disabilities, as compared with lifetime best, was 6.88 (SD = 1.76; MIN = 2; MAX = 10). In the group of non-working people with acquired motor disabilities, the corresponding value was 4.37 (SD = 2.41, MIN = 0, MAX = 10). t-Tests for independent samples revealed that the mean work ability, as compared with the values corresponding to lifetime best, was a significantly differentiating factor for the studied groups (p = 0.00).

The current work ability with respect to the physical and mental demands was rated as "rather good" by 42.4% and 40.8%, respectively, of the participants with acquired motor disabilities. According to 39.2% of the respondents, the disabilities do not exclude job performance, but it causes some symptoms. Due to health problems, 32.8% of the respondents were absent from work from 10 to 24 days during the past year, whereas 53.6% were not sure whether they would be able to do their current jobs within two years since then.

Most of the answers to the question "Have you recently been able to enjoy your regular daily activities?" were "rather often". Accordingly, most of the answers to the question "Have you recently been active and alert?" were also "rather often". The answers to the question "Have you recently felt yourself to be full of hope for the future?" were most often "rather often", "often" or "sometimes".

WORK ABILITY PREDICTORS IN WORKING INDIVIDUALS WITH ACQUIRED MOTOR DISABILITIES

Stepwise regression analysis of work ability was applied in working persons with acquired motor disabilities, where work ability was a dependent variable. In stepwise regression analysis, only the variables that are significant predictors of the dependent variable can be added to the model. According to the rules for adding the variables into the model, continuous and dichotomous variables were selected.

TABLE 15.1

Summary of Work Ability Model in Working Persons with Motor Acquired Disabilities

R	R-Squared	Corrected R-Squared	Standard Error
0.573	0.328	0.308	4.44275

TABLE 15.2

Factors Significantly Affecting Work Ability in Working Individuals with Acquired Motor Disabilities

Model	B	Standard Error	Beta	t	Significance
Constant	30.120	2.913		10.340	0.000
Disease acceptance	0.242	0.059	0.264	4.100	0.000
Adjustment of workstation to the needs resulting from disability	1.671	0.573	0.156	2.915	0.004
Level of optimism	0.247	0.086	0.177	2.879	0.004
Job matching the learned profession	−1.573	0.585	−.147	−2.691	0.008
Age	−0.116	0.030	−0.231	−3.822	0.000
Duration of living with a disability	0.130	0.047	0.169	2.754	0.006
Workplace	−1.778	0.660	−0.156	−2.694	0.008

The model obtained from the stepwise regression analysis explains about 31% of dependent variable variance (Table 15.1).

Out of all the considered variables, seven significantly affecting work ability were selected for the model (Table 15.2).

The results obtained from working participants with acquired disabilities having an optimistic outlook and accepting their disability, living with a disability within a longer period of time and working in the environment adapted to their requirements, revealed higher self-rating of work ability in this group. Conversely, lower self-rating of work ability was noted in older respondents working beyond their areas of expertise or in protected work environment.

WORK ABILITY PREDICTORS IN NON-WORKING INDIVIDUALS WITH ACQUIRED MOTOR DISABILITIES

As in the case of the working participants, stepwise regression analysis was conducted to determine work ability in the non-working individuals with acquired disabilities. Self-rating of the current ability to do any job, as compared with the lifetime best was an independent variable. According to the rules of adding variables to the model, continuous and dichotomous variables were selected.

The model obtained from the stepwise regression analysis explains about 48% of dependent variable variance (Table 15.3).

TABLE 15.3

A Summary of Current Work Ability Rating in Non-working Individuals with Acquired Motor Disabilities

R	R-squared	Corrected R-squared	Standard Error
0.691	0.477	0.464	1.77270

TABLE 15.4

Factors Significantly Affecting Current Work Ability Rating in Non-working Individuals with Acquired Motor Disabilities

Model	B	Standard Error	Beta	t	Significance
Constant	2.736	0.653		4.192	0.000
Health status rating	−1.919	0.254	−0.393	−7.568	0.000
Disability acceptance	0.109	0.015	0.364	7.028	0.000
Employer's reluctancy	0.877	0.260	0.159	3.378	0.001
Gender	−0.620	0.229	−0.128	−2.706	0.007
Children	0.580	0.241	0.114	2.406	0.017
Education	−0.812	0.386	−0.100	−2.105	0.036

Out of all the studied variables, six significantly affecting work ability were finally added to the model (Table 15.4).

The results obtained from the non-working individuals with acquired motor disabilities, who accept their conditions, have children and report the employer's reluctance as the reason of their unemployment, revealed higher work ability rating. Conversely, the females rating their health status as poor and those with lower education levels gave lower scores to their work ability.

As gender turned out to be a differentiating factor in work ability assessment, stepwise regression analysis was performed to determine predictors of work ability for both sexes.

Non-working Women

The model obtained from stepwise regression analysis of the non-working women's and men's parameters explains 44% of and about 49% of dependent variable variance, respectively (Table 15.5).

Among all the considered variables significantly affecting work ability, five were included in the female model and two were included in the male model (Table 15.5).

Older women reporting poor health, getting no support or encouragement from their closest friends or families, tend to underestimate their work ability. Conversely, the women who have accepted their disabilities and report employer's reluctancy

TABLE 15.5

Summary of the Model Assessment of Current Work Ability and the Factors Significantly Affecting Assessment of Current Work Ability in Non-working Women and Men with Acquired Motor Disabilities

Women				
R	R-squared	Corrected R-squared		Standard Error
0.680[h]	0.463	0.440		1.62994
Model	B	Standard Error	Beta	t	Significance

Model	B	Standard Error	Beta	t	Significance
Constant	3.214	0.917		3.505	0.001
Health status assessment	−1.344	0.334	−0.306	−4.028	0.000
Disability acceptance	0.117	0.022	0.379	5.243	0.000
Employer's reluctance	1.254	0.351	0.245	3.573	0.001
Age	−0.036	0.014	−0.190	−2.651	0.009
Support and encouragement from families and friends	−1.306	0.509	−0.178	−2.567	0.012

Men				
R	R-squared	Corrected R-squared		Standard Error
0.704b	0.496	0.487		1.87747
Model	B	Standard Error	Beta	t	

Model	B	Standard Error	Beta	t	
Constant	3.532	0.670		5.274	•0.000
Health status assessment	−2.446	0.386	−0.466	−6.341	•0.000
Disability acceptance	0.103	0.021	0.353	4.802	•0.00

as the cause of unemployment (possibly resulting from delegating responsibility to the employers or thinking: "I am able to work, but this is the employer who doesn't want me") were found to attribute higher scores to their work ability.

Conversely, the male respondents who rated their health status as poor attributed lower scores to their work ability, while higher scores were obtained from men who accepted their disability.

DISCUSSION

Becoming a disabled person entails negative emotions, frustrations and stress, and imposes new challenges on that person, such as the desire to overcome the resultant obstacles (Trzebińska 2007). Work ability assessment plays a key role in reinforcement of the worker's self-esteem, self-confidence or independence. It also affects the decision whether to engage in occupational activity or not. Thus, based on work ability assessment we can predict whether a person with a disability will return to work or will be excluded from the labour market (Johansson et al. 2006).

As regards self-rating of work ability, 57% of the working respondents with acquired disabilities rated it as moderate. According to WAI scale, the score obtained from this assessment was 32.8 points. Persons with disabilities reporting good work ability significantly more often are willing to work and enter the labour market as compared with those who believe that their work ability is impaired. The latter are more often unemployed or perform a monotonous work, physical work or work beyond their capacity (Boman et al. 2015).

The study results indicate that the participants perceive their disability as a factor limiting their work ability. According to McGonagle et al. (2015), the perceived work ability is the predictor of such phenomena as absenteeism, retirement or leaving work due to disability. Lavasani et al. (2015) in turn indicate that work ability in people with disabilities strongly and positively correlates with such factors as job satisfaction and core evaluation encompassing fundamental evaluations of themselves, their abilities, self-confidence and a positive attitude to themselves and the surrounding world.

Optimistic employers with acquired motor disabilities who have accepted their disability live longer with their disability, and their working environment is adjusted to their requirements usually report higher levels of work ability. Older persons, in turn, who don't do their acquired jobs and are working in protected work environment tend to underestimate their work ability.

Research results show that work ability decreases with age and the pace of this decrease depends on such factors as health status or the type of job performed (Makowiec-Dąbrowska et al. 2008). The age-related decrease in work ability is confirmed by many other studies (Sjögren-Rönkä et al. 2002; van den Berg et al. 2009; El Fassi et al. 2013; de Vries et al. 2013); there are also studies determining that the 45th year of age is the cut-off age for the unemployed (Hult et al. 2018). Another view on the relation between age and work ability assessment has been presented by Cochrane et al. (2018), reporting higher work ability ratings in older participants.

Moreover, hard work, repeatability of activities within the working cycle and uncomfortable body positions, as well as long working hours are the factors responsible for premature complete loss of work ability. The aforementioned factors are supposed to contribute to work ability reduction in working individuals (Makowiec-Dąbrowska et al. 2008). Gamperiene et al. (2008) has found a correlation between low work ability and doing a job requiring no skills. Physically demanding work is also the factor responsible for a decrease in WAI (El Fassi et al. 2013; Oliv et al. 2017). In case of workers with acquired motor disabilities, meeting their needs, e.g., through workplace adaptation is the factor significantly contributing to the improvement of their work ability (Brouer et al. 2010). Researchers believe that physical adjustment of the working environment is an effective form of instrumental social support. It increases the worker's chance to return to both part-time and full-time job.

The non-working individuals who have accepted their disability, have children and report the employer's reluctance to hire them as the reason of unemployment ("I am able to work, this is the employer who doesn't want to hire me") gave higher scores to their work ability. The lower rating of work ability as well as the reasons for not getting hired results from health status assessment (poor health), gender (female) and education levels. The persons reporting poor self-rated health are likely to self-rate their work ability as low, unlike those reporting good health (Kosskinen 2008). The persons with chronic nonspecific musculoskeletal pain reporting good self-rated health and convinced that they can effectively relieve their pain tend to rate their work ability higher than those having the same condition, with poor self-rated health, believing that they are unable to effectively relieve their pain (de Vries et al. 2013). Additionally, work ability can be decreased due to high-intensity musculo-skeletal symptoms (Sjögren-Rönkä et al. 2002). A low potential for musculoskeletal pain treatment is also related to lower values of WAI (van den Berg et al. 2008), while lower levels of functional limitations correlate with higher corresponding values (Cochrane et al. 2018).

The effect of illness acceptance on work ability level, in both working and non-working individuals, is of note. Acceptance of illness impacts multiple areas of patients' lives. It is closely related to such issues as the perceived quality of life, activity and therapy outcome. It is the basis of disease perception, providing opportunities for deriving satisfaction from past events and hope for the future, or giving up in case of no chance for overcoming the difficulties and feeling helpless. Acceptance of illness is a determinant of functioning with it. It increases confidence in physicians and treatment approaches (Ślusarska et al. 2016). Each patient or person with a disability has his/her own conception of illness and thus its cognitive reflection. It is the basis of cognitive assessment, defining the core emotions associated with the illness and in consequence, influencing the decisions what actions should be undertaken, including those related to changes in one's health status (Heszen and Sęk 2007). The feeling of helplessness and the resulting low level of acceptance result in deterioration not only of the quality of life but also of the level of professional, cognitive and social functioning (Ślusarska et al. 2016). Fear and work avoidance are the most important factors responsible for the unwillingness to undertake occupational activity 3 months and one year after acquiring a disability. Almost 50% of variance corresponding to such behaviours is explained by subjective sensation of health-related discomfort, disease perception and education level (Øyeflaten et al. 2008). There is no doubt that disease acceptance plays a pivotal role in coping with everyday problems experienced by people with and disabilities. It is responsible for the effectiveness and outcomes of rehabilitation processes and social inclusion of persons with disabilities and is closely related to work ability self-assessment.

REFERENCES

Bishop, M. 2005. Quality of life and psychosocial adaptation to chronic illness and acquired disability: A conceptual and theoretical synthesis. *J Rehabil* 71(2):5–13.

Boman, T., A. Kjellberg, B. Danermark, and E. Boman. 2015. Employment opportunities for persons with different types of disability. *Alter* 9(2):116–129.

Brouer, S., M. F. Reneman, U. Bültmann, J. J. van der Klink, and J. W. Groothoff. 2010. A prospective study of return to work cross health conditions: Percived work attitude, self-efficacy and percived social support. *J Occup Rehabil* 20(1):104–112.

Brzezińska, A. I., R. Kaczan, K. Piotrowski, and P. Rycielski. 2008. Uwarunkowania aktywności zawodowej osób z ograniczeniami sprawności: czynniki wspomagające i czynniki ryzyka. *Nauka* 3:97–123.

Brzezińska, A. I., R. Kaczan, and P. Rycielski. 2010. Model czynników warunkujących wykluczanie/ inkluzję osób z ograniczeniami sprawności na rynku pracy. *Polityka Społeczna*, Nr specjalny: Diagnoza potrzeb i podstawy interwencji społecznych na rzecz osób z ograniczeniami sprawności, 43–45.

Bugajska, J. 2018. *Model oceny zdolności do pracy dla potrzeb aktywizacji zawodowej młodych osób z niepełnosprawnością ruchową: Podręcznik i procedury.* Warszawa: CIOP-PIB.

Bugajska, J., A. Jędryka-Góral, R. Gasik, and D. Żołnierczyk–Zreda. 2011. Nabyte zespoły dysfunkcji układu mięśniowo-szkieletowego u pracowników w świetle badań epidemiologicznych. *Med Pr* 62(2):153–161.

Byra, S. 2014. Nadzieja podstawowa i percepcja własnej niepełnosprawności a radzenie sobie osób z urazem rdzenia kręgowego. *Hygeia Public Health* 49(4):825–832.

Cochrane, A., N. M. Higgins, C. Rothwell et al. 2018. Work outcomes in patients who stay at work despite musculoskeletal pain. *J Occup Rehabil* 28(3):559–567.

de Vries, H. J., M. F. Reneman, J. W. Groothoff, J. H. B. Geertzen, and S. Brouwer. 2013. Self-reported work ability and work performance in workers with chronic nonspecific musculoskeletal pain. *J Occup Rehabil* 23(1):1–10.

DeSanto-Madeya, S. 2009. Adaptation to spinal cord injury for families post-injury. *Nurs Sci Q* 22(1):57–66.

El Fassi, M., V. Bocquet, N. Majery, M. L. Lair, S. Couffignal, and P. Mairiaux. 2013. Work ability assessment in a worker population: comparison and determinants of Work Ability Index and Work Ability score. *BMC Public Health* 13:305.

Gamperiene, M., J. F. Nygård, I. Sandanger, B. Lau, and D. Bruusgaard. 2008. Self-reported work ability of Norwegian women in relation to physical and mental health, and to the work environment. *J Occup Med Toxicol* 3:8.

Ge, H., X. Sun, J. Liu, and C. Zhang. 2018. The status of musculoskeletal disorders and its influence on the working ability of oil workers in Xinjiang, China. *Int J Environ Res Public Health* 15(5):842.

Heszen, I., and H. Sęk. 2007. *Psychologia zdrowia*. Warszawa: Wydawnictwo Naukowe PWN.

Hult, H., A. M. Pietilä, P. Koponen, and T. Saaranena. 2018. Association between good work ability and health behaviours among unemployed: A cross-sectional survey. *Appl Nurs Res* 43:86–92.

Johansson, G., O. Lundberg, and I. Lundberg. 2006. Return to work and adjustment latitude among employees on long-term sickness absence. *J Occup Rehabil* 16:185–195.

Juczyński, Z. 2012. *Narzędzia pomiaru w promocji i psychologii zdrowia*. Wyd. 2. Warszawa: Pracownia Testów Psychologicznych Polskiego Towarzystwa Psychologicznego.

Koskinen, S., T. Martelin, P. Sainio, and R. Gould. 2008. Factors affecting work ability: Health. In *Dimensions of work ability: Results of the Health 2000 Survey*, eds. R. Gould, J. Ilmarinen, J. Järvisalo, and S. Koskine. Helsinki: Helsinki Finnish Centre for Pension.

Lavasani, S. S., N. A. Wahat, and A. Ortega. 2015. Work ability of employees with disabilities in Malaysia: Disability. *CBR Inclusive Dev* 26(2):22–46.

Lidal, I. B., T. K. Huynh, and F. Bierung-Sørensen. 2007. Return to work following spinal cord injury: A review. *Disabil Rehabil* 29(17):1341–1375.

Lipińska-Lokś, J. 2007. Osoby z niepełnosprawnością wobec konieczności nagłej zmiany roli zawodowej. In *Osoby z niepełnosprawnościami na współczesnym rynku pracy*, eds. B. Pietrulewicz, and M. A. Paszkowicz, 21–28. Zielona Góra: Instytut Edukacji Techniczno-Informatycznej. Uniwersytet Zielonogórski.

Majewski, T. 2011. *Poradnictwo zawodowe i pośrednictwo pracy dla osób niepełnosprawnych. Poradnik dla urzędów pracy*. Warszawa: Ministerstwo Pracy i Polityki Społecznej. Biuro Pełnomocnika Rządu do Spraw Osób Niepełnosprawnych.

Makowiec-Dąbrowska, T., W. Koszada-Włodarczyk, A. Bortkiewicz et al. 2008. Zawodowe i pozazawodowe determinanty zdolności do pracy. *Med Pr* 59(1):9–24.

McGonagle, A. K., G. G. Fisher, J. L. Barnes-Farrell, and J. W. Grosch. 2015. Individual and work factors related to percived work ability and labor force outcomes. *J Appl Psychol* 100(2):376–398.

Miranda, H., L. Kaila-Kangas, M. Heliövaara et al. 2010. Musculoskeletal pain at multiple sites and its effects on work ability in a general working population. *Occup Environ Med* 67:434–435.

Oliv, S., A. Noor, E. Gustafsson, and M. Hagberg. 2017. A lower level of physically demanding work is associated with excellent work ability in men and women with neck pain in different age groups. *Saf Health Work* 8(4):356–363.

Olney, M. F., K. Brockelman, J. Kennedy, and M. A. Newsom. 2004. Do you have a disability? A population-based test of acceptance, denial, and adjustments among adults with disabilities in the U.S. *J Rehabil* 70(1):4–9.

Ottomanelli, L., and L. Lind. 2009. Review of critical factors related to employment after spinal cord injury: Implications for research and vocational services. *J Spinal Cord Med* 32(5):503–531.

Øyeflaten, I., M. Hysing, and H. R. Eriksen. 2008. Prognostic factors associated with return to work following multidisciplinary vocational rehabilitation. *J Rehabil Med* 40(7):548–554.

Pawłowska-Cyprysiak, K. 2011. Uwarunkowania jakości życia osób z niepełnosprawnością ruchową. *Bezpieczeństwo Pracy – Nauka i Praktyka* 10:6–8.

Pawłowska-Cyprysiak, K. 2015. Reintegracja zawodowa osób z niepełnosprawnością ruchową nabytą w trakcie kariery zawodowej. *Niepełnosprawność – zagadnienia, problemy, rozwiązania* 4(17):45–56.

Pawłowska-Cyprysiak, K., M. Konarska, and D. Żołnierczyk-Zreda. 2013a. Self-perceived quality of life of people with physical disabilities and labour force participation. *Int J Occup Saf Ergon* 19(2): 185–194.

Pawłowska-Cyprysiak, K., M. Konarska, and D. Żołnierczyk-Zreda. 2013b. Uwarunkowania jakości życia osób niepełnosprawnych ruchowo. *Med Pr* 64(2): 227–237.

Powdthavee, N. 2009. What happens to people before and after disability? Focusing effects, lead effects, and adaptation to different areas of life. *Soc Sci Med* 69(12):1834–1844.

Saunders, S. L., and B. Nedelec. 2014. What work means to people with work disability: A scoping review. *J Occup Rehabil* 24(1):100–110.

Sjögren-Rönkä, T., M. T. Ojanen, E. K. Leskinen, S. T. Mustalampi, and E. A. Mälkiä. 2002. Physical and psychosocial prerequisites of functioning in relation to work ability and general subjective well-being among office workers. *Scand J Work Environ Health.* 28(3):184–190.

Ślusarska, B., G. J. Nowicki, M. Serwata et al. 2016. Poziom akceptacji choroby i jakość życia chorych na chłoniaki. *Med Paliat* 8(2):88–95.

Trzebińska, E. 2007. Integracja społeczna w ujęciu psychologicznym. In *Oddziaływania psychologiczne na rzecz integracji osób z ograniczeniami sprawności*, ed. E. Trzebińska. Warszawa: Wydawnictwo Naukowe Academica SWPS/EFS.

Tuomi, K., J. Ilmarinen, A. Jahkola, L. Katajarinne, and A. Tulkki. 1994. Work ability index. In *Occupational Health Care*, ed. S. Rautaoja. Helsinki: Institute of occupational Health.

Tuomi, K., L. Eskelinen, J. Toikkanen, E. Jarvinen, J. Ilmarinen, and M. Klockars. 1991. Work load and individual factors affecting work ability among aging municipal employees. *Scand J Work Environ Health* 17(1):128–134.

Van den Berg, T. I., S. M. Alavinia, F. J. Bredt, D. Lindeboom, L. A. Elders, and A. Burdorf. 2008. The influence of psychosocial factors at work and life style on health and work ability among professional workers. *Int Arch Occup and Environ Health* 81(8):1029–1036.

Van Leeuwen, C. M., M. W. Post, T. Hoekstra et al. 2011. Trajectories in the course of life satisfaction after spinal cord injury: Identification and predictors. *Arch Phys Med Rehabil* 92(2):207–213.

Vornholt, K., P. Villotti, B. Muschalla et al. 2018. Disability and employment – Overview and highlights. *Eur J Work Org Psychol.* 27(1):40–55.

Wainwright, E., D. Wainwright, E. Keogh, and C. Eccleston. 2013. Return to work with chronic pain: Employers' and employees' views. *Occup Med* 63:501–506.

Wolski, P. 2010. *Utrata sprawności: Radzenie sobie z niepełnosprawnością nabytą a akty-wizacja zawodowa.* Warszawa: Wydawnictwo Naukowe SCHOLAR.
Wolski, P. 2013. *Niepełnosprawność ruchowa: Między diagnozą a działaniem.* Warszawa: Centrum Rozwoju Zasobów Ludzkich.
Zawieska, W. M. 2014. *Przystosowanie obiektów, pomieszczeń oraz przystosowanie stanow-isk pracy dla osób niepełnosprawnych o specyficznych potrzebach: Ramowe wytyczne.* Warszawa: CIOP-PIB.

Index

Printed in the United States
by Baker & Taylor Publisher Services